T0135746

I dedicate this thesis to
Reinhard Heinrich
and Judith Lyons.

Computational biology of microRNAs and siRNAs

Dissertation
Zur Erlangung des akademischen Grades
doctor rerum naturalium (Dr. rer. Nat)

im Fach Biologie eingereicht an der
Mathematisch-Naturwissenschaftlichen Fakultät I
der Humboldt Universität zu Berlin

von
Debora Susan Marks BSc.
geboren am 27.09.1957, Leeds, England

Präsident der Humboldt-Universität zu Berlin:
Prof. Dr. Dr. h.c. Christoph Markschies

Dekan der Mathematisch-Naturwissenschaftlichen Fakultät I:
Dekan Prof. Dr. Andreas Herrmann

Gutachter

1. Professor Dr. Martin Vingron
   Director at the Max Planck Institute for Molecular Genetics,
   Max Planck Institute for Molecular Genetics Berlin

2. Professor Dr. Burkhard Rost
   Department for Bioinformatics and Computational Biology
   Fakultät für Informatik, Technische Univeristät, München

3. Professor Dr. Peter Hammerstein
   Mathematisch-Naturwissenschaftliche Fakultät I,
   Humboldt-Universität

Datum der Promotion: 30. Juni 2010

Keywords
Bioinformatics, Systems Biology, microRNA, siRNA, RNAi

Bibliographic information published by the Deutsche Nationalbibliothek

The Deutsche Nationalbibliothek lists this publication in the Deutsche
Nationalbibliografie; detailed bibliographic data are available in the
Internet at http://dnb.d-nb.de .

Logos Verlag Berlin GmbH
Comeniushof, Gubener Str. 47
10243 Berlin
Tel.: +49 (0)30 42 85 10 90
Fax: +49 (0)30 42 85 10 92
INTERNET: http://www.logos-verlag.de

0   Abstract

0.1   English abstract

Thousands of microRNAs, ~22 nucleotide RNAs, ~5% of the gene content in all metazoans, regulate the dose of protein coding genes by imperfect complimentary matching to the mRNA, mediating degradation of mRNA and reducing translation. The first part of this thesis (Chapters 2 and 3) describes a flexible computational method to identify microRNA target genes (miRanda, www.microrna.org) in flies, fish and mammals, which guided focused experiments and predicted the scope of regulation. miRanda predicted that a large fraction of expressed genes were under microRNA control, particularly transcription factors and RNA binding proteins and that genes were under combinatorial targeting, reminiscent of transcription factor regulation. Although hundreds of miRanda target predictions have now been experimentally verified, many of the observed gene expression changes, after microRNA and siRNA perturbations, remained unexplained. The second part of this thesis (Chapter 5 and 6) takes a more *system* level and quantitative approach to the problem of the sensitivity of microRNA and siRNA targeting. The thesis identifies strong experimental support for the hypothesis that saturation of required protein machinery, RNA-induced Silencing Complex (RISC) and competition between mRNAs for microRNAs (and siRNAs) could drive the level of down-regulation, in addition to the specifics of the binding sites. Hundreds of published genome-wide expression experiments in many cell types supported that genes with endogenous microRNA sites were upregulated, plausibly due to saturation of the machinery. Similarly, these expression experiments also supported the idea that target transcript abundance dilutes microRNA and siRNA efficacy. *Global* cellular properties affected the *individual* microRNA: mRNA (and siRNA:mRNA) activity. These system level properties of the cell are vital for a fuller understanding of microRNA cellular function, siRNA design and therapeutic development.

0.2   Zusammenfassung

Tausende von microRNAs, kleine RNA-Molekülen von ca. 22 Nukleotiden Länge, die zusammen bis zu 5% des Genrepertoires aller mehrzelligen Eukaryonten ausmachen, regulieren die Quantität proteinkodierender Gene durch Bindung an mRNAs, und reduzieren damit die Translation von Gen zu Protein. Diese Arbeit beschreibt im ersten Teil eine vielseitige, computerbasierte Methode (miRanda, www.microrna.org), diese Target-Gene von microRNAs in Fliegen, Fischen und Säugern vorherzusagen. Auf Basis von miRanda-Vorhersagen wurden bereits gerichtete Experimente durchgeführt und das Maß der microRNA-induzierten Regulation abgeschätzt. miRanda sagte vorher, dass ein hoher Anteil exprimierter Gene, vor allem Transkriptionsfaktoren und RNA-binding Proteine, unter der Kontrolle von microRNAs steht, und dass Gene sich unter dem Einfluss von kombinatorischem ‚targeting' befinden, mit ‚Gedächtnis' an die entsprechende Regulation durch Transkriptionsfaktoren.

Obwohl mittlerweile Hunderte von miRanda-Vorhersagen experimentell verifiziert wurden, sind doch viele der beobachteten Änderungen der Genexpression, nach Störung durch microRNAs oder siRNAs, bisher unerklärt.Im zweiten Teil dieser Arbeit wird mittels eines mehr systembiologischen und quantitativen Ansatzes das Problem der Sensitivität des microRNA- und siRNA-targeting adressiert. Es werden starke experimentelle Argumente für die Hypothese präsentiert, dass eine Art Sättigung der relevanten Protein-Maschinerie, RNA-induced Silencing Complex (RISC) und ein Wettbewerb zwischen mRNAs um die entsprechenden microRNAs and siRNAs den Grad der (down-) Regulierung bestimmt, in Ergänzung zur Spezifizität der jeweiligen Bindungsstellen.

Hunderte von genomweiten Expressionsexperimenten in vielen verschiedenen Zelltypen haben belegt, dass Gene mit endogenen microRNA-Loci hochreguliert wurden, höchstwahrscheinlich durch Sättigung der jeweiligen Proteinmaschinerie. In ähnlicher Weise lieferten diese Experimente auch weitere Anhaltspunkte, dass ein Überfluss an Target-Transkripten die Wirksamkeit von microRNAs/siRNAs reduziert. *Globale* zelluläre Zustände beeinflussten die *individuelle* Aktivität von microRNA bzw. siRNA gegenüber mRNA. Das Verständnis dieser systemischen Eigenschaften der Zelle sind grundlegend wichtig für ein volles Verständnis sowohl der zellulären Funktion von microRNAs, als auch des siRNA-Design und entsprechender therapeutischer Ansätze.

1   Chapter One –
Introduction to microRNA biology

1.1   Ideas are rarely new

The idea that RNA-RNA duplexes might regulate the expression of genes is not new. In 1961, Jacob and Monod elucidated their hypothesis of messenger RNA in a review paper, Genetic Regulatory Mechanisms in the Synthesis of Proteins (Jacob and Monod 1961), in which they suggested that certain RNA messages may repress genes by repression of the DNA directly or repression of the RNA "structural message" which encodes the protein, Figure1.1.

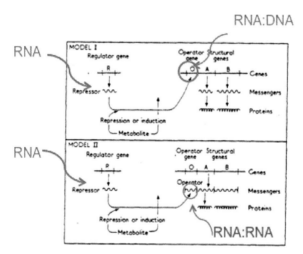

Figure 1.1 Two models of gene regulation presented by Jacob and Monod in 1961 (Jacob and Monod 1961) *Blue arrows point to proposed RNA, RNA:DNA or RNA:RNA interactions they proposed as mechanisms for gene repression, my annotation.*

They went on to win the Nobel prize for the discovery of mRNA in 1965. Although, subsequently, it was found that the lac repressor was in fact a protein, nevertheless Jacob and Monod clearly saw the theoretical possibility of RNA-DNS and RNA-RNA regulation in their models.

In the main, this hypothesis that RNA:RNA interactions mediated gene regulation, was forgotten until the work of Roy Britten and Eric Davidson in the seventies. In a series of papers they reported the discovery of nuclear RNA, which did not code for proteins or any other known functional RNA. Indeed they found that sea urchin blastula, Hela cells were dominated by these smallish RNAs (Britten and Davidson 1969; Britten and Davidson 1971; Davidson and Britten 1979). They postulated a theory around "Possible Mechanisms of Processing Control by RNA:RNA duplexes" (Davidson and Britten 1979), Figure 1.2. RNA:RNA duplexes which could regulate gene expression almost disappeared from the mocleular biological consciousness until the discovery of siRNAs and microRNAs some 40 years later.

Figure 1.2 Elements of the Regulation model, reproduced from Britten and Davidson 1979. *Annotation to show RNA:RNA duplex controlling gene expression*

1.2   What are microRNAs?

1.2.1   Early discoveries

microRNAs are small (~22 nucleotides) RNAs that mediate post-transcriptional silencing of genes by perfect or imperfect base pairing with target mRNAs .

In1993 the Ruvkun and Ambros labs simultaneously reported a small RNA called *lin-4*, Figure 1.3, the founding member of the miRNA family in *C. elegans*. They discovered it through a genetic screen for defects in the temporal control of post-embryonic development (Lee, Feinbaum et al. 1993; Wightman, Ha et al. 1993). lin-4 mutations disrupt the temporal regulation of larval development in one of the 4 larval stages in *C. elegans*. By reinteraing the first stage (L1) at a later time in development. Worms that are deficient for the lin-4 target protein coding gene lin-14 on the other hand, display opposite developmental phenotypes. Most

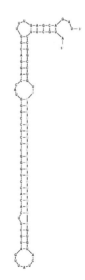

Figure 1.3 C. elegans lin-4 hairpin *Pre-mature lin-4 folded using Mfold (Zuker 2003)*

genes identified from mutagenesis screens up to then had been protein-coding, so it came as quite a surprise to those in the two research groups that this small 22-nt RNA could be responsible for this crucial developmental control. They discovered that the micorrna lin-4 regulates lin-14 by imperfect complimentary binding to 7 sites in the 3'-untranslated region (UTR), Figure 1.4

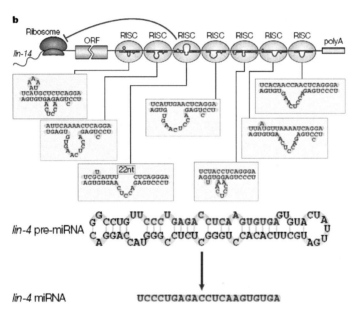

Figure 1.4 The targeting of lin-14 by the microRNA lin-4 a. *the target sites in the 3'UTR proposed by Ambros 1993. B. the stem loop of lin-4 calculated with RNA fold. Image reproduced from Hannon review (He and Hannon 2004)*

It took another 7 years before the Ruvkun lab reported the second microRNA, *let-7* (Reinhart, Slack et al. 2000). let-7 was also discovered genetically and encodes a 21-nucleotide small RNA that controls the developmental transition from the L4 stage into the adult stage. Similar to lin-4, let-7 functions by binding to a the 3' UTR, this time of of lin-41 and hbl-1 (lin-57), and inhibiting their translation. The mechanism of of lin-4 and let-7 were thought to be of translational inhibition alone, with the mRNA stalling on the polysomes and no detectable effect on mRNA levels. Since that time there have been a number of landmarks studies showing that the mRNA can indeed be down-regulated, as well as the protein, though the exact correlation and mechanism remains under dispute (Filipowicz, Bhattacharyya et al. 2008; Bartel 2009). The identification of let-7 not only provided another example of developmental regulation by small RNAs, but also raised the possibility that such RNAs might be present in species other than nematodes. let-7 and its target lin-41 were clearly evolutionary conserved throughout metazoans with homologs easily detected in mollusks, sea urchins, flies, mice and humans (Pasquinelli, Reinhart et al. 2000). (This was unlike lin-4, whose vertebrate homlog, miR-125, escaped notice for another 2 years). The extensive conservation indicated a more general role of small RNAs in developmental regulation.

The discovery of let-7 was quickly followed in 2001 by a screen for more small RNAs by three teams, led by Ambros, Tuschl and Bartel (Lagos-Quintana, Rauhut et al. 2001; Lau, Lim et al. 2001; Lee and Ambros 2001). They cloned 21-25-nt RNAs from *Drosophila*, mouse, worm and human, identifying ~ 100 new microRNAs, some conserved in all four species, such as let-7, miR-1 and many clustered together in the genome suggesting that they were transcribed together, Figure 1.5. Following this Science "Discovery of the Year", the biological community responded over the next 5 years with a slew of experimental and bioinformatic methods designed to identify new microRNAs in many model organism, enabled by the newly published genome sequences in human, rat, mouse, rat, fish, *Drosophila*, worm, and now some 13 other mammals. (Berezikov, Cuppen et al. 2006). The number of article in PubMed on microRNAs is still increasing exponentially

My thesis starts with work begun in 2002 and I report the development of a bioinformatic pipeline which used experimentally cloned sequences form different stages of *Drosophila* and fish development to identify new candidate microRNAs, Chapter 2.

In 2004 the Tuschl lab searched for a viral siRNAs as a response in human cells to viral infection. To their surprise, they instead discovered the first viral microRNAs in Epstein Barr Virus (HHV4) infected B cells, (Pfeffer, Zavolan et al. 2004); towards the end of Chapter Three, I describe my role in that work and the prediction of viral microRNA targets in the host, and in the virus itself. Viral microRNAs were also discovered in other large DNA viruses, including other gamma herpes families ,(Cai, Lu et al. 2005; Pfeffer, Sewer et al. 2005; Cullen 2006; Gottwein, Mukherjee et al. 2007; Ganem and Ziegelbauer 2008). The evolutionary relationship of viral and human microRNAs remains an open question, though mir-155 has a homolog in the Epstein Barr virus genome (Gottwein,

Mukherjee et al. 2007), it is not clear how many others are functionally related. I explore this further in Chapter 3.

1.1.2   Biogenesis

microRNAs are bound to Argonaute proteins in the complex RNA-induced Silencing Complex (RISC), where they mediate mRNA degradation and inhibition of translation initiation of mRNAs, for extensive reviews of the microRNA biogenesis see (Tomari and Zamore 2005; Filipowicz, Bhattacharyya et al. 2008; Ghildiyal and Zamore 2009). Briefly, microRNAs are typically transcribed by RNA polymerase II and are often polycistronic, Figure 1.5, and have a typical exon – intron structure of protein coding genes. The sequence of events during microRNA biogenesis is important to understand for the latter part of this thesis, where I suggest rate limiting steps involving protein complexes, may be saturated in certain circumstances, leading to global shifts in which genes are expressed and by how much.

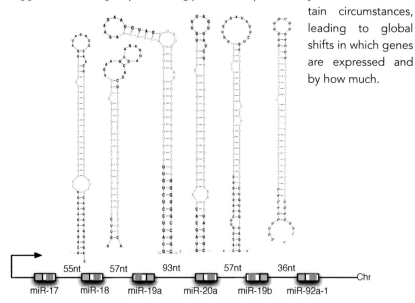

Figure 1.5 Example of polycistronic human microRNAs *Fold-back secondary structures involving miRNAs (green) and their flanking sequences (black), as predicted computationally using RNAfold (Hofacker and Stadler 2006) a. microRNAs identified in a polycistronic cluster b. Schematic of distance. Green main strand expressed, dotted green, " star" strand.*

The biogenesis (after transcription) begins with the excision of a pri-microRNA hairpin, which is cleaved from the transcript by Drosha and DGCR8, and exported as the pre-microRNA from the nucleus by exportin-5 and a Ran GTP system. Once in the cytoplasm both Dicer and Argonaute can cleave the hairpin to form the microRNA duplex (Cheloufi, Dos Santos et al. ; Cifuentes, Xue et al.) which is the unwound. One, or both strands, is separately loaded into AGO1 or AGO2, the complex called RISC, see Figure 1.4. and binds to mRNA, causing a reduction in protein translation. The situation is more complicated than that of course. In fact, microRNAs in a cluster can be separately transcribed, the 'microprocessor' complex conatinig Drosha and Pasha has some site flexibility (Seitz and Zamore 2006) and the microRNA transcripts can be edited by ADAR (Yang, Chendrimada et al. 2006). Feedback regulation may also occur in some states, a target of let-7, lin-28, can inhibit let-7 biogenesis, causing double negative feedback implying a possible switch like mechanism (Rybak, Fuchs et al. 2008). Finally recent work suggests that under conditions of cellular stress and starvation, microRNAs are able to up-regulate target mRNAs (Vasudevan, Tong et al. 2007; Steitz and Vasudevan 2009), but the extent of this dual role is still under investigation.

1.1.3   microRNAs continue to be discovered

After the initial wave of microRNA identification about 250 more miRNAs were estimated to be encoded in the human genome, but it was subsequently recognized that this estimate could be low (Berezikov, Thuemmler et al. 2006)v. Later studies, based on combinations of computational and experimental techniques, supported a now confirmed, substantially larger number of miRNAs. (www.mirbase.org) It is still difficult to estimate the upper limit of miRNAs in humans and other mammals, in part because of the difficulty in defining what a 'true' miRNA is. In 2003, an *ad hoc* group from the microRNA community agreed on a list of criteria to be use in various combinations as acceptable (Ambros, Bartel et al. 2003). In summary these were:

i.   Detection of a distinct 22-nt RNA transcript by hybridization to a size-fractionated RNA sample (ordinarily by the Northern blotting method).

ii.   Identification of the 22-nt sequence in a library of cDNAs made from size-fractionated RNA. Such sequences must precisely match the genomic sequence of the organism from which they were cloned (except as noted below).

iii.   Prediction of a potential fold-back precursor structure that contains the 22-nt miRNA sequence within one arm of the hairpin. In this criterion, the hairpin must be the folding alternative with the lowest free energy, and must include at least 16 bp involving the first 22 nt of the miRNA and the other arm of the hairpin. In animals, these fold-back precursors are usually about 60–80 nt, whereas in plants, they are more variable, and may include up to a few hundred nucleotides.

iv.   Phylogenetic conservation of the 22-nt miRNA sequence and its predicted fold-back precursor secondary structure. The conserved hairpin should meet the same minimal pairing requirements as in criterion C, but need not be the lowest free energy folding alternative.

v.   Detection of increased precursor accumulation in organisms with reduced Dicer function

The group suggested various combinations of the criteria, which would be sufficient to confirm a candidate microRNA as true. In reality, the criteria used has varied from group to group and over time. The use of phylogentic conservation is problematic in two senses: first the use of this as a criterion automatically disqualifies microRNAs that are species specific and secondly it begs the question of *how much* conservation is enough.

Reviews of microRNA cloning methods show some convergence and optimisation of methods (Tagami, Inaba et al.). In 2007 Landgraf et al., sequenced over 250 small RNA libraries from 26 different organ systems and cell types of human and rodents(Landgraf, Rusu et al. 2007). This massive compendium clearly displays the wide variation in tissue and cell type specificity of microRNA expression in mammals. This profiling includes many hematopoietic malignancies allowing a direct comparison to the normal tissues. As well as being a resource, the compendium highlights the expression of just a few microRNAs is enough to describe most the differences between cell types and tissues.

New high sensitive sequencing technologies ("deep sequencing") for microRNAs can now capture tens of millions of small RNAs from a single libraries (Hafner, Landgraf et al. 2008)5, allowing the discovery of hundreds of cell specific microRNAs expressed at concentrations too low to detect by more conventional cloning techniques. These deep sequencing technologies have identified many new microRNAs in specific locations and temporal expression, such as the female reproductive tract (Creighton, Benham et al.), lipocyte development (Hicks, Trakooljul et al.), human embryonic stems cells (Koh, Sheng et al.) and cancers such as prostate (Szczyrba, Loprich et al.) melanoma (Stark, Tyagi et al.)

ovarian (Wyman, Parkin et al. 2009). In addition it has been used to discover microRNAs in animals such as the Tasmaninan Devil (Murchison, Tovar et al.) and different species of Drosophila (Lu, Fu et al. 2008; Lu, Shen et al. 2008), as well as a number of viral infected human cells (Umbach, Wang et al.; Umbach, Kramer et al. 2008; Umbach, Nagel et al. 2009), using methods to determine probabilities for authentic microRNA existence(Emde, Grunert et al.; Friedlander, Chen et al. 2008) .

Will deep sequencing of microRNAs be important to understand the functions of microRNAs in an organism? When applied even to a single HeLa cell, deep sequencing can recover 40% of all human microRNAs as compared to recovery on only ~ 5% with conventional cloning. Whilst this may be important, it is not yet clear how "functional' microRNAs, expressed at very low concentrations, actually are. However, some evidence suggests atht at least some are: microRNAs with very limited spatial expression, such as lsy-6 in the neural circuitry of the C. elegans, can have profound cell fate consequences. Interestingly, my own calculations suggest an inverse relationship between number of gene targets for a specific microRNA and its level of expression, suggesting an fucntional and evolutionary relationship somwhwat explore in work on mciroRNAs in Drosophila species (Lu, Shen et al. 2008).

By 2009, 10,000 microRNAs had been identified across a wide range of metazoans, some single cell eukaryotes, and in plants, see miRBase http://www.mirbase.org/, Figure 1.6b. In humans alone there are over 1400 annotated. In human cells there may be as many as 500,000 and as few as 10,000 individual microRNA molecules in others. (My estimation based on extrapolation from the calculation done in C. elegans and supplementary material from Hannon lab (He, He et al. 2007). Whatever the exact number of microRNAs in a cell, it seems that they may be just one order of magnitude less than the number of mRNA molecules themselves, see BioNumbers (http://bionumbers.hms.harvard.edu/).

Perhaps it is therefore not so surprising that the number of PubMed indexed studies on microRNAs is still increasing exponentially, reaching 2,400 in the period January- October 2009 alone, and speaks to the massive impact of microRNAs on the whole breadth of biological and clinical research, Figure 1.6a

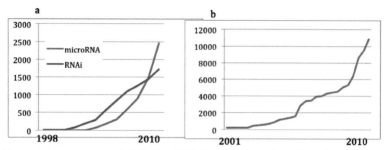

Figure 1.6 The exponential of microRNA discovery. *Number of a. PubMed articles on microRNA and RNAi from 1998-2010 b. number of microRNAs in miRBase*

1.3   RNAi discovery and siRNAs in functional genomics

Two years before the discovery of the second miRNA, let-7, small RNAs were also found in a seemingly distinct biological process, RNA interference (RNAi). RNAi is an evolutionarily conserved, sequence-specific gene-silencing mechanism that is induced by exposure to dsRNA. Andrew Fire and Craig Mello went on to win the 2006 Nobel Prize for their discovery published in 1998 that double stranded RNA could be harnessed by cells in in C. elegans to silence genes, which are complementary to one of the strands.

Their work finally illuminated the puzzle of co-suppression in plant biology that had remained unexplained for a decade. In the 1980s plant researchers encountered two unexpected complications with antisense technology. First, most of the plant lines with an antisense transgene did not exhibit suppression of the corresponding endogenous gene (Smith et al., 1988). Second, some of the control lines with the sense constructs exhibited coordinate suppression (cosuppression) of the transgene and the homologous endogenous gene (Napoli et al., 1990; Smith et al., 1990;van der Krol et al., 1990). One of the quintessential co-suppression experiments involved attempting to increase the colour intensity of the petals in petunias. However, the surprise was that the introduction of a red pigment gene led in fact to colour *loss* and white petals.

A similar puzzle had arisen in C. *elegans* research. Gene expression could be specifically suppressed by direct injection of antisense RNA; however, the process was inefficient because large amounts of antisense RNA had to be injected (Fire et al., 1991). An inspired analysis of the C. *elegans* phenomenon provided the key to understanding these unexpected findings. It revealed that RNAi in injected *C. elegans* was mediated by a small amount of double-stranded (ds) RNA that contaminated the sense and antisense RNA preparations (Montgomery et al., 1998). Its was this agent, the dsRNA which caused the gene silencing. A second key discovery followed this, fist disovered in plants, was that the specific agents involved were antisense species of ~25nts processed from the longer dsRNA strands (Hamilton Baulcombe 1999). The following year this was confirmed in flies with the concomitant discovery of the RISC. An avalanche of discoveries was precipitated, for review see XX.

We now now that a dsRNA of ~500 bp is the stimulus to initiate RNAi in many organisms including worms, plants and flies. The long dsRNA is ultimately processed in vivo into small dsRNAs of ~21–25 bp in length, designated as small interfering RNAs (siRNAs). The sense strand of the siRNA duplex is selectively incorporated into an effector complex (the RNA-induced silencing complex; RISC). The RISC directs the cleavage of complementary mRNA targets, a process that is also known as post-transcriptional gene silencing (PTGS). In mammals and other vertebrates the successful use of RNAi depends on the more direct introduction of preprocessed siRNAs to avoid the interferon response. The design of these siRNAs and eventually shRNAs involved some important chemical modifications and understating of the natural RNAi requirements in cells for a 3'overhang and 5' phosphate group to ensure efficient loading into RISC.

This astounding development of the ability to silence genes at the whim of the scientific investigator, really once only the domain of yeast biologists, has opened up serious whole genome functional experiments with si and sh libraire designed to target specific genes of interest or genome wide screens for specific effects. Importantly it has also led to the development of clinical applications with many siRNA products now in stage there clinical trials .

In the Chapters 4 and 5 of thesis, I examine some of the general consequences of siRNA transfections of cells and how that might impinge both functional genomics and the therapeutic developments.

1.4   MicroRNA target prediction

The ~22 nucleotide mature microRNA only partially matches the targeted mRNA, and the exact configurations "allowed" are still under debate. As described in section 1.2, the first binding sites identified between the microrna lin-4 and mRNA, lin-14 contained 6-8 matches in the 5' end of the mciroRNA and some atching in the 3' end. This partial matching means that there can be tens of thousands of possible matches for each microRNA in the 3'UTRs of animal genomes. These predictions are typically calculated assuming only 3'UTR targeting and not, for instance, the coding region, so using current understanding of microRNA targeting, this would be an order more sites in the genome with some microRNAs targeting as many as one third of expressed mRNAs in a cell.

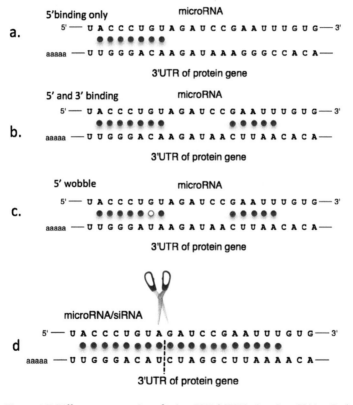

Figure 1.7 Different categories of microRNA/siRNA sites in mRNAs. *Red dot G:C match, Blue to A:U match open red, G:U wobble a. So called 'seed only site' with perfect matching from positions 2-8 b. Binidng at both ends So-called "canonical" site c. Wobble (or mismatch) in 5' end, so called "compensatory" site. d. Perfect matching past position 12, microRNA will behave like an siRNA and cut the mRNA between positions 9 and 10 of the mi/siRNA.*

From early on in microRNA research, target sites were categorized into three types (Brennecke, Stark et al. 2005)s (i) 5' matching, positions 2-9 no 3' matching ("seed" only (Bartel 2009) (ii) 5' exact matching and some 3' matching (iii) 5' mismatch and 3' matching (sometimes called "compensatory" (Bartel 2009) or "non-canonical" (Brennecke, Stark et al. 2005), see Figure 1.7. These different categories are all thought to mediate some mRNA degradation to different extents and different research groups claim different relative importance of the different sites. Some work suggest that there are few "effective" non-canonical binding sites (Hafner, Landthaler et al. ; Bartel 2009), though others and myself have argue that the assessment of whether this is the case is inadequate (Vella, Reinert et al. 2004; Didiano and Hobert 2008; Khan, Betel et al. 2009)

The main target prediction methods used by the community are outline in Table 1.1 (This is not meant to be an exhaustive list, but to contain those programs which account for mmore than 80% usage of community usage. This is my estimate based on prediction program citations). Initially, microRNA target predictions were validated in two ways,; one through shuffling of UTRs, and second, an anecdotal method, through small numbers of luciferase reporter experiments. These latter experiments were simple overexpresed luciferase reporter systems, with sites to match microRNAs placed in the 3'UTR. Dampening of fluorescence was taken as validation of the microRNA regulation, with additional validation from rescue mutations of the binding sites (Lewis, Shih et al. 2003). This method of validation is limited since it may not reflect in vivo regulation of the gene by the specific microRNA, since concentrations of both the reporter and microRNA may be far from physiological (Didiano and Hobert 2006; Didiano and Hobert 2008).

In 2005 Lim et al., showed that despite previous dogma, mRNA levels do indeed change in a measurable way after microRNAs are over expressed in cell culture. Since that time microarray measured mRNA gene expression changes have been the benchmarks by which target prediction methods validate their algorithms, and produce false positives and negatives (Lewis, Burge et al. 2005; Nielsen, Shomron et al. 2007). The most widely used target prediction programs do not allow any mismatch or G:U wobble in the positions 2-8 (Krek, Grun et al. 2005; Grimson, Farh et al. 2007). Although sites with mismatches are thought to be in a small minority (< 10%) (Hafner, Landthaler et al. ; Bartel 2009), this is now clearly at odds with known, functional microRNA binding sites

(Zisoulis, Lovci et al. ; Didiano and Hobert 2006; Tavazoie, Alarcon et al. 2008). In addition, most microRNA target prediction methods are limited by reliance upon over expression experiments followed by mRNA measurements to validate computational methods.

When individual microRNA sites are discovered in a non biased way, genetically, or new methods are used which may be *less perturbing* (for example, AGO pulldown, or microRNA inhibition) a different picture emerges (Zisoulis, Lovci et al.) (also my evidence, not shown).

Indeed extensive detailed experiments investigating the exact binding requirements with more realistic, *in vivo* readouts, reveals a much more subtle picture of functional microRNA binding sites (Vella, Reinert et al. 2004; Didiano and Hobert 2006; Didiano and Hobert 2008). In one of these studies, the Hobert lab shows that three of the five constructs containing G:U wobbles in both target sites displayed regulation, albeit at levels below wild-type regulation These results indicate that G:U wobbles can be detrimental to *lsy-6*-mediated regulation in specific circumstances; however, G:U wobbles should not be used as a criterion by which to eliminate potential miRNA/target interactions. Other newer work in *C. elegans* shows the tolerance of G:U wobbles for function sites (Hammell; Hammell, Long et al. 2008). The extent of mismatching in the 5′ end which can be tolerate for a functional site, remains an open question.

Since 2003 there have been hundreds of attempts to predict microRNA targets computationally. miRanda, presented at the beginning of this thesis was the first target prediction method published (Enright, John et al. 2003) (October 2003). This was contemporaneous with a method by Cohen group in Drosphila (Stark, Brennecke et al. 2003) followed closely by TargetScan (Lewis, Shih et al. 2003). Then followed by PicTar (Rajewsky and Socci 2004) and Diana g (Kiriakidou, Nelson et al. 2004) the following year. PicTar, can to all intents and purposes be regarded a equivalent to TargetScan. Most of the predictions between the two programs overlap and the constraints were similar. The differences in the predictions may have as much to do with the different sets of UTRs used, and different conservation rules as they do with any substantive differences in the programs (Ritchie, Flamant et al. 2009). Table 1.1 provides a brief summary of the methods presented chronologically.

Program : PMID	Publication year	Matching criteria	Energy calculation	Conservation	Scoring scheme	Download	online search
miRanda (Enright, John et al 2003; John, Sander et al. 2006)	www.microrna. org	Weighted alignment emphasizing positions 2-8	Hybridization energy threshold	AVID alignment of UTRs	Alignment score, multiple sites additive	Yes	yes
TargetScanS (Lewis, Shih et al. 2003; Friedman, Farh et al. 2009)	www.targetscan. org	Match in position 2-8 required	None	Whole genome alignments	Two scoring methods, one with conservation, one with context	No	No
Diana-microT (Kiriakidou, Nelson et al. 2004)	diana. pcbi. upenn. edu	Dynamic programming - loops	Hybridisation energy threshold + rules	Not implemented	All equal	No	Yes
PicTar (Rajewsky and Socci 2004)	pictar. mdc-berlin.de	Seedmatch or compensatory	Hybridization energy threshold (RNAfold).	Whole genome alignments	Maximum likelihood,	No	No
RNAhybrid (Rehmsmeier, Steffen et al. 2004)	bibiserv. techfak. uni-bielefeld.de/ rnahybrid	Integrated match and energy	Guarantees to find lowest energy site	Not implemented	extreme value dist. BLAST-like evalue	no	yes

Rna22 (Miranda, Huynh et al. 2006)	cbcsrv. watson. ibm. com/ rna22. html	Best align-ment	None	'Target Islands'	extreme value dist. BLAST-like evalue	No	Yes
MicroInspec-tor (Rusinov, Baev et al. 2005)	bioinfo. uni-plo-vdiv.bg/ microin-specto	Perfect seed, fol-lowed by alignment	Hybrid-ization energy threshold (RNAfold)	Not imple-ment-ed	Sort by energy	No	Yes
miTarget (Kim, Nam et al. 2006)	cbit.snu. ac.kr/ ~miTar-get	Short seed	Hybrid-ization energy threshold (RNAfold)	Not imple-ment-ed	SVM that weight a combina-tion of features learned from ex-amples	No	Yes
PITA (Kertesz, Iovino et al. 2007)	genie. weiz-mann. ac.il	6-mers; wobbles allowed	Filtered by net energy of binding	Not imple-ment-ed	ΔΔG	no	no
miRWIP (Hammell, Long et al. 2008)	146.189. 76.171	6-mers, *C. elegans* only	Secondary structure	Not imple-ment-ed	heuristic	no	no

Most tools in this table rely on overexpression experiments for either training or validation or both. Exceptions are miRWIP, which uses AGO pull down manual inspection criteria, RNAHybrid, Diana-T and miRanda which were developed before the genome wide experiments showed mRNA changes in expression after microRNA overexpression (Lim, Lau et al. 2005)a. In the early days of tar-get rpediction many assumed that the duplex energy calculation would be an important predictor of the level of repression on an mRNA (Enright, John et al. 2003; Rehmsmeier, Steffen et al. 2004), but this proved to be only very marginally predictive alone, or suffered from >90% false negative rate. One reason for this might be that the microRNA is held by Argonaute and is therefore already over some entropic barriers in some postions, so weaker matching might be possible. Later methods such as PITA and miRWIP calculate the net gain in energy formation, accounting for loss of energy from mRNA:mRNA binding. The jury is still out on the quality of these methods. The three target prediction methods most widely used (miRanda, TragetScan and PicTar) produce a large number of false positives (as measured by numbers of genes with target sites which do not change in expression after overexpression of a microRNA). This is discussed further in Chapter 3 and is the motivation for the development of the approach taken in Chapters 4 and 5.

1.5   Functions of microRNAs

The earliest experiments individual functions of the hundreds of microRNAs may be quite distinct, as indicated by their specific spatial and temporal expression (Lagos-Quintana, Rauhut et al. 2001; Sempere, Sokol et al. 2003). However the question what the overlal function of microRNAs was in development of the organism. Dicer knock-outs in C. elegans were embryonic lethal, and Dicer knock-outs in zebrafish suggested that microRNAs were important for morphogenesis but necessarily differentiation itself. Howvere the zebrafish developmental defects after Dicer knock out developed no further than X (Giraldez, Cinalli et al. 2005). It is beyond the scope of this thesis to summarize the thousands of "individual" functions ascribed to microRNAs and I recommend the interested reader to start with various reviews (function reviews). However, some themes have emerged. Cell-type specific microRNAs often target genes important for muscle proliferation and differentiation eg miR-133 and miR-1 in muscle m which are co-trancribed and necessary for for muscle proliferation and differentiation (Chen, Mandel et al. 2006), miR-124 expressed in developing neurons, involved in neuroogenesis in human and mouse, synaptic learning in *aplysia* (Rajasethupathy, Fiumara et al. 2009) miR-122, highly expressed

in liver, regulates lipid metabolism (Krutzfeldt, Rajewsky et al. 2005; Esau, Davis et al. 2006), cholesterol biosynthesis (Elmen, Lindow et al. 2008), may help defend against hepatitis C (Jopling, Yi et al. 2005). microRNAs as a whole tend to be down-regulated across many different types of cancers (Lu, Getz et al. 2005), and can be used in groups to classify different cancers in a way which has proved difficult using any other groups of genes (Lu, Getz et al. 2005). Human microRNA genes are also frequently located at fragile sites and genomic regions involved in cancers (Calin, Sevignani et al. 2004). For example, one copy of the polycistronic miR-15 and miR-16 microRNAs is located in *dleu2*, a transcript deleted in some leukemias and for which no protein product had been found (Calin, Dumitru et al. 2002). Similarly let-7 is thought to have a tumour suppressor role in lung cancer (Esquela-Kerscher, Trang et al. 2008). Nevertheless, some microRNAs are said to be 'oncogenic', for instance miR-19 from the miR-17-5p cluster (Mu, Han et al. 2009), miR-10b breast cancer metastisis (Ma, Teruya-Feldstein et al. 2007) and miR-155, originally identified as the BIC gene, can accelerate the pathogenesis of lymphomas and leukemias *(Eis, Tam et al. 2005)*. For a fuller discussion see Chapter 4.

1.6   Global effects on proteins by microRNAs

Recent work showed that protein levels are altered by microRNA overexpression (Baek, Villen et al. 2008; Selbach, Schwanhausser et al. 2008). This is hardly surprising since this was the original premise (Lee, Feinbaum et al. 1993). The question was how much correlation there is between the predicted target genes and the protein changes. Both the Selbach et al., and Baek et al., use their own target gene definitions. Both reports show that genes with target sites in their mRNAs are significantly more likely to be reduced in protein expression than all other genes. However, the correlations between changes in mRNAs and proteins is not as strong as might have been expected, and at least 40% of genes which are downregulated at the mRNA level, with sites are not at protein level and vice versa, see Chapter 5. In addition the changes in protein levels is quite low, and others have commented on whether the reported conclusions of the Selbach et al. work are justified from the data (Margolin, Ong et al. 2009).

1.7   Systems biology of microRNA targeting – a quantitative approach

Most previous work on microRNA target prediction defines whether a gene is a target or not, rather than a more quantitative approach which is more likely to reflect the biology. For instance predictions should ideally have a probability associated with them, with the amount of change in gene expression being part of the prediction. Methods have concentrated in the specifics of the binding site, and honed on changes in gene expression at the mRNA level after microRNA over expression with transfections. These changes may be far from physiological targeting, so we await new kinds of data such as Ago pull downs using the clip methods, or a larger number of microRNA inhibtion experiments. Also, ideally we would like to see more high through put protein measurements. The second part of this thesis, Chapters 4 and 5, starts to address the issue of a more quantitative approach, by considering global properties of the cell which may affect the amount of gene regulation by a microRNA or siRNA. The summary in Chapter 6 contains some examples of future directions which the work could take, and outlines in some detail how this approach could be developed.

2   Chapter Two – microRNA discovery and target gene prediction

2.1   Synopsis

This chapter first briefly describes building the bioinformatic pipeline for microRNA discovery in *Drosophila melanogaster*, followed by the substantive part of the Chapter on the development of the miRanda algorithm for microRNA target prediction, and third, the prediction of microRNA function in *D. melanogaster* using microRNA target predictions. In particular, the last part highlights the prediction of combinatorial regulation of target genes with a many to one and a one to many relationship, with a significant proportion of the genome under microRNA control.

Published papers which contain some of the work in this Chapter:

Primary
MicroRNA targets in *Drosophila*. Genome Biol. 2003;5(1):R1. 2003 Dec 12 PubMed ID:14709173. Enright AJ, John B, Gaul U, Tuschl T, Sander C, **Marks DS**. (Conceived project, designed algorithm, designed computational validation, analysed data, wrote the paper)

Secondary
The small RNA profile during *Drosophila* melanogaster development. Dev Cell. 2003 Aug;5(2):337-50. PubMed ID: 12919683, Aravin AA, Lagos-Quintana M, Yalcin A, Zavolan M, **Marks D**, Snyder B, Gaasterland T, Meyer J, Tuschl T. (Bioinformatic pipeline design for microRNA discovery, data analysis)

2.2   microRNA gene discovery by cloning and bioinformatics

2.2.1   Cloning

The primary method of microRNA discovery started with the cloning of size fractionated RNAs followed by a bioinformatics pipeline. The work described in this chapter is the development of a bioinformatic pipeline for some of the first *Drosophila* microRNAs which were discovered after the initial reports in Science in 2001 (see Chapter 1 for discussion) The cloning was done by the Tuschl laboratory. Briefly, it involved extracting isolating small RNAs from RNA from cell lysate. RNAs of the desired size range (~19 to 25 nt in mammals) were gel-isolated from total RNA and subjected to sequential 3'- and 5'-adapter ligation; adapters allow the priming reverse transcription and definition of the orientation of the cloned small RNA, followed by sequencing, (Lagos-Quintana, Rauhut et al. 2001; Lagos-Quintana, Rauhut et al. 2002). Total RNA was isolated from embryos at 5 different timepoints after egg laying as well as from larval stages 1, 2, and 3 stages, from pupae, adults, and microdissected adult testes. Subsequent to predicted microRNA discovery after analysis of the cloned sequences, predicted microRNAs were tested using Northern blots.

2.2.2   What was the computational challenge?

Potential microRNAs were predicted from the concatenated sequences produced from sequencing of the clones. Crucially, sequences from other genomic functional units needed to be identified and excluded, such as mRNA, tRNA, ribosomal RNA and other RNA fragments not excluded in the small RNA cloning methodology.

Clearly this also raised the issue of the definition of a microRNA itself; we used the definition provided by the consortium, as discussed in Chapter One (Ambros, Bartel et al. 2003) The calculations that followed determined the likelihood that the sequence discovered was a bone fide microRNA.

2.2.3   Computational pipeline

A bioinformatics pipeline was designed to input cloned sequences and output predicted microRNAs. Cloned sequences were compared to all known genome annotation by using a variety of databases including the annotation of the *D. melanogaster* genome (version 3.1 from http://www.bdgp.org), a dataset of *D. melanogaster* sequences from GenBank,

(http://www.fruitfly.org/sequence/sequence_db/na_gb.dros), a database of transposable elements
(http://www.bdgp.org/p_disrupt/datasets/VERSION3/ALL_SEQUENCES_
dmel_RELEASE3.FASTA.ALL.v3) and canonical sequences
(http://www.bdgp.org/p_disrupt/datasets/NATURAL_TRANSPOSABLE_ELE-
MENTS.fa), a database of *D. melanogaster* tRNA sequences
(http://rna.wustl.edu/GtRDB/Dm/Dm-seqs.html), a database of small RNA

sequences provided by A. Hüttenhofer, and a database of miRNAs (http://www.sanger.ac.uk/Software/Rfam/ftp.shtml) and predicted miRNA sequences (Lim et al., 2003).

The assignment of annotation was performed in a hierarchical manner. Perfect matches of small RNA sequences to the genomes of S. cerevisiae were classified as S. cerevisiae sequences followed by matches to *D. melanogaster* rRNA, tRNA, and then snRNAs/snoRNAs and other ncRNAs. The remaining sequences were then checked against euchromatic and heterochromatic *D. melanogaster* genomic sequences as well as GenBank sequences of other organisms and classified as mRNAs, *Drosophila* C Virus (DCV), bacterial, and plant genomes. The annotation for small RNAs as tRNA breakdown products was performed using the tRNA sequences provided at the Genomic tRNA Database at http://rna.wustl.edu/tRNAdb/ as well as annotated GenBank sequences. snRNA and snoRNA hits were assigned by using the sequence sets provided in (Yuan et al., 2003). The noncoding RNAs (ncRNAs) were provided from A. Hüttenhofer and contained longer RNA sequences (>40 nt) without annotation or assigned function. After annotation of a fraction of the sequences, the remainder were tested.

2.2.4   Classes of *D. melanogaster* small RNAs

A total of 4074 clone sequences were obtained and current public databases were used to annotate 95.6% of these sequences; the residual sequences could not be annotated because they did not match to any of the sequenced genomes in the database (183 clones) or because they matched to a region of the *D. melanogaster* genome for which no functional or sequence homology assignment could be made (40 clones). The largest class of cloned RNAs represents breakdown products of abundant noncoding (or nonmessenger) RNAs (rRNA, tRNA, snRNA, snoRNA, and others) of *D. melanogaster* (63.5%), followed by breakdown product sequences from *Saccharomyces cerevisiae* (12.9%), which constitutes the preferred diet of *D. melanogaster*. A few bacterial and plant rRNA fragments from salmonella, cereal, and hops were also found, presumably because the baking yeast that was used for fly food was brewing yeast. A small fraction of *D. melanogaster* mRNA breakdown products (3.5%) was also identified, see Table 2.1 for details. The residual 577 sequences (14.1%) fell into the following three classes: miRNAs (9.3%), repeat-associated small RNAs derived from sense and antisense strands of repetitive elements (4.4%), and small RNAs from *Drosophila* virus C (0.4%).

Type	Embryo Stage (hr)					Larva Stages			Pupa	Adult	Testes	Total (clones)	Total (%)
	0-2	2-4	4-6	6-12	12-24	L1	L1 + L2	L3					
rRNA	161	119	160	85	23	193	111	293	91	438	451	2160	53.0
tRNA[a]	25	14	25	1	2	35	64	53	9	99	25	354	8.7
miRNA	112	42	25	7	3	12	8	8	—	121	41	382	9.4
rasiRNA	79	39	13		1	2	3	—	—	12	28	178	4.4
mRNA	3	7	11	2	1	26	30	32	1	16	13	143	3.5
snRNA/snoRNA	4	—	4	2	3	1	2	2	—	16	1	35	0.9
Other ncRNA	8	1	7	1	—	1	3	1	1	13	3	40	1.0
S. cerevisiae	45	101	49	22	19	53	61	26	1	141	7	525	12.9
Bacteria, plants	1	1	2	—	—	—	6	4	1	2	—	17	0.4
DCV	—	—	—	—	—	1	1	11	1	3	—	17	0.4
Unknown[d]	31	14	7	2	7	15	32	25	8	56	25	223	5.5
Total	469	338	303	122	59	339	321	455	113	920	594	4074	100

Table 2.1. Composition of Small RNA cDNA Libraries Prepared from Different Developmental Stages and Testes of D. melanoga

The proportion of small RNAs with regulatory function relative to rRNA break-down products varied drastically for the different developmental stages. Early embryos and adults showed the highest content of small regulatory RNAs (between 15% and 40%) while in late embryo and larvae the number was significantly lower. From pupae, almost all cloned small RNAs represented rRNA and tRNA breakdown products. Presumably, extensive apoptosis during metamorphic tissue reorganization caused extensive rRNA breakdown and made it impossible to clone small RNAs at high frequency. Also, we found that pupae- and adult-specific miRNAs, such as let-7, were still present in pupae RNA preparations and were readily detectable by Northern blotting.

2.2.5   microRNA discovery

After excluding rRNA and tRNA other sequences were taken through the bioinformatic pipeline as described above. According to the convention for miRNA annotation (Ambros, Bartel et al. 2003), new miRNA gene names were assigned based on the evidence of cloning the small RNA from the cDNA libraries and their phylogenetic conservation as a fold-back precursor structure in other species. In total we identified 382 clones that were derived from 60 miRNA gene, Table 2.1. Some of the cloned miRNAs, such as miR-1 through miR-14, let-7, and bantam were described previously (Pasquinelli, Reinhart et al. 2000; Lagos-Quintana, Rauhut et al. 2001; Lau, Lim et al. 2001; Lee and Ambros 2001; Brennecke, Hipfner et al. 2003). miRNAs that closely resemble in sequence previously described microRNAs were miR-9b, 9c, 31a, 31b, 34, 79, 92a, 92b, 124, 184, and 210. These were named based on their evolutionary relationships to pre-identified miRNAs. About 70% of the identified sequences of miRNAs are between 21 and 23 nt in size with an average of 22.0 nt This size distribution is similar to the length distribution of siRNAs generated by in vitro processing of dsRNA in D. melanogaster embryo lysate (Elbashir, Martinez et al. 2001). For some of the microRNAs we also cloned the strand opposite to the accumulating and conserved miRNA. We refered to these sequences as the miR* sequence (Lau, Lim et al. 2001) . In most cases, miR* is clearly less abundant, but for miR-10*, miR-13a*, and miR-281-2*, too few sequences were cloned to make any conclusion on the relative abundance. A similar situation was found for two small RNAs that are excised from a hairpin residing in the noncoding RNA transcript iab-4 of the Bithorax complex. iab-4 contributes to proper formation of abdominal segments (Mattick and Gagen 2001) mirna genes are often found in close proximity to each other forming larger mirna gene clusters. We found 11 gene clusters in the *D. melanogaster* genome containing on average three miRNAs, with the longest cluster containing eight miRNAs, Supplementary Table 2.1. Some clusters are found within intergenic regions, while others are located within the intronic regions of protein-coding genes. Because positional clustering of miRNAs is a common genomic feature of miRNAs, we examined the regions adjacent to mirna genes for the presence of additional miRNAs that may have escaped the nonsaturating cloning and sequencing protocol. By identifying fold-back structures as well as sequence homologs of known miRNAs and only considering those candidates that were also conserved in other insect genomes, we identified eight more miRNAs. These include mir-283 clustered with mir-12 and mir-304, mir-100 clustered with let-7 and mir-125, mir-313 and mir-310 clustered with mir-311 and mir-312, and mir-2c clustered with mir-13a and mir-13b. Nonclustered conserved fold-back structures were identified for mir-87 and mir-133. The expression of miR-87, -100, -125, -133, -283, -310, and -312 was confirmed by Northern analysis. The validation of the predicted miR-2c was not attempted because of predictable problems of cross-hybridization of the Northern blotting probe to miR-2a and miR-2b. To date, including the predicted clustered and/or conserved miRNAs, a total of 62 unique mirna sequences encoded by 68 genes have been identified in *D. melanogaster*.

Although we did not clone miR-125, the lin-4 homolog of fly, we previously showed that mir-125 was expressed in pupa and adult (Lagos-Quintana, Rauhut et al. 2002). As expected from their close proximity, the expression pattern of mir-125 was identical to that of *D. melanogaster* let-7 (Pasquinelli, Reinhart et al. 2000). A third mirna gene that is homologous to mammalian mir-100 was found in close proximity to mir-125 and let-7, Figure 2.1. We confirmed that *D. melanogaster* miR-100 is coexpressed with miR-125 and let-7 by Northern blotting. Coexpression of miR-125 and let-7 and induction by the steroid hormone ecdysone at the onset of metamorphosis has been reported recently Bashirullah et al. 2003 and Sempere et al. 2003, as well as the coregulated expression of miR-100 (Bashirullah, Pasquinelli et al. 2003; Sempere, Sokol et al. 2003). The polycistronic expression of this mirna cluster was also confirmed by the Tuschl lab by RT-PCR analysis, which detected a long primary transcript comprising all the hairpin precursor sequences.

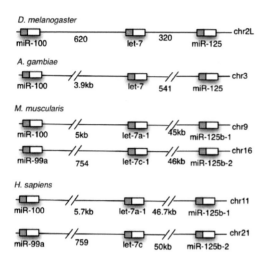

Figure 2.1 Arrangement of the miRNA genes in *D. melanogaster*. *The 70 nt fold-back precursor is indicated as box, the position of the miRNA in the precursor is shown in black. The chromosome location is indicated to the right. The miRNA clustering is conserved between invertebrates and vertebrates but gene duplication occurred in mammals and the spacing between miRNA precursors increased with increasing genome size.*

The clustering for mir-100, mir-125, and let-7 is conserved in *A. gambiae*, although the distance between the genes has increased, Figure 2.1 In mouse and human, the gene cluster underwent duplication and the distance between miRNA genes was increased further, Figure 2.1. In *C. elegans*, neither lin-4 nor its paralog mir-237 are spatially linked with let-7 family members (Lim, Lau et al. 2003). Evidence supporting the conservation of lin-4 and let-7 coregulation in nematodes comes from the identification of both lin-4 and let-7 RNA binding sites in the 3' UTR of lin-14, lin-28, lin-41, and lin-57 transcripts (Reinhart and Ruvkun 2001; Abrahante, Daul et al. 2003). Homologs to miR-100 were not identified in *C. elegans*. The 5' portion of the miRNAs shows the highest degree of conservation and miRNAs can be grouped in families based on this criterion Ambros et al. 2003b and Lim et al. 2003b. It was also noticed that the 5' ends of some miRNAs are complementary to sequence motifs in the 3' UTR of some mRNA that are subject to posttranscriptional regulation. The 62 unique miRNA sequences can be grouped into 43 families, 14 of which are universally conserved, Supplementary Table 2.2 – my table. For three families, the conservation extends to mammals only, and for six families the conservation extends to nematodes only. Notably, 17 small RNAs matched the *Drosophila C virus RNA.* We isolated 17 small RNAs from *DCV*, 16 of which were in + orientation and one in – orientation. *Drosophila C virus (DCV)* belongs to the class of *Dicistroviridae* and the genus *Cripavirus* (Cricket paralysis-like virus) and contains a positive-sense genomic RNA of 9264 nucleotides in length (Johnson and Christian, 1998).

Gene	Mature miRNA and miRNA* Sequence	Size Range (nt)	Chr	Position, Orientation	Embryo (hr)					Larva			A	T	Total
					0-2	2-4	4-6	6-12	12-24	1	1+2	3			
mir-303	UUUAGUUUCACAGGAAACUGGU	23	X	4106936...58,–										1	1
mir-31b	UGGCAAGAUGUCGGCAUAGCUG	22	X	8732382...403,–									1		1
mir-13b-2§	UAUCACAGCCAUUUUGACGAGU	17-24	X	8890259...80,+	3					1					4
mir-283*	UAAAUUCAGCUGGUAAUUCU	21*	X	15208989...9008,+										4	7
mir-304	UAAUCUCAAUUUGUAAAUGUGAG	21-23	X	15239984...40006,+	2							2	2	3	13
mir-12	UGAGUAUUUACAUCAGGUUACUGGU	21-24	X	15240403...515,+	5	1	3						3		3
mir-210	UUGUGCGUGUGACAGCGGCGU	20-23	2L	17859250...49,+	1	1									6
mir-275	UCAGGUACUUGAAGUAGUGGCGCG	19-25	2L	7416084...106,+	1			1					1		2
mir-305	AUUGUACUUCAUCAGGUGCUCUG	19-23	2L	7418210...32,+	4	2				1			5		12
mir-2b-1§	UAUCACAGCCAGCUUUGAGGAGCG	19-24	2L	8250853...75,–											
mir-87*†	UGAGCAAAUUUCAGGUGUG	21	2L	9942846...65,–											9
mir-9c	UCUUUGGUAUUCUAGCUGUAGA	22	2L	16576020...41,+	1	1							1	1	2
mir-305	UCAGGUACUUGGUGACUCUCA	22	2L	16576502...22,+											1
	*CGGCGUCACUCUGUGGCGUGGC	22	2L	16576539...60,+											2
mir-79	UAAAGCUAGAUUACCAAAGCAU	22	2L	16576697...718,+	1	1	2			2			1		6
mir-9b	UCUUUGGUGAUUUUAUGCUGUAUG	21-23	2L	16576842...64,+											5
mir-124†	UAAGGCACGCGGUGAAUGCCAAG	19-23	2L	17544509...31,+											
mir-100b*	AACCCGUAAAUCCGAACUUGUGG	22	2L	18449528...49,+											
let-7*	UGAGGUAGUAGGUUGUAUAGU	20-21	2L	18450127...147,+									3	2	5
mir-125*n	UCCCUGAGACCCUAACUUGUGA	22	2L	18450426...47,+	3	3	2			1			4	2	15
mir-2a-2	UAUCACAGCCAGCUUUGAUGAGCU	20-24	2L	18547578...601,–	2	3	2						2	2	4
mir-2a-1*	*CCUCAUCAAGUGGUUGUGAUA	21	2L	18547617...37,–	3	3				1			4	2	15
mir-2b-2§	UAUCACAGCCAGCUUUGAGGAGCGG	19-24	2L	18547984...8007,–	4	2	2				2		5	2	12
mir-1†	UGGAAUGUAAAGAAGUAUGGAG	20-27	2L	16548271...94,–	23	1	4			4		3	34	3	74
mir-133*n	UUGGUCCCCUUCAACCAGCUGU	22	2L	20586415...36,+											4
mir-14	UCAGUCUUUUUCUCUCUCCUA	19-25	2R	484434...54,+	2		2						1		4
mir-307	UCACAACCUCCUUGAGUGAG	20	2R	4681996...915,–									1		1
mir-281-2§	UGUCAUGGAAUUGCUCUCUUUGU	23	2R	7234976...98,–						1			1		1
mir-281-1§	*AAGAGAGCUAUCCGUCGACACU	22	2R	7234922...43,–						1			1		
mir-184	UGUCAUGGAAUUGCUCUCUUUGU	23	2R	7235096...118,–	5	2	1	1	1				12	4	26
	UGGACGGAGAACUGAUAAGGGC	18-23	2R	8394134...55,–								1	1		1
mir-308	*CCUUAUCAUUCUCUCGCCCCG	21	2R	8394174...94,–	4	1							1		6
mir-279*	AAUCAGAGAUAUUACUGUGAG	18-22	2R	9279528...49,–						1			1	6	8
mir-9*	UCGUGGGACUUUCGUCCGUUU	20-22	2R	10720847...68,+								1			1
	*AUGUUACGGCGAGGAGUUAGA	21	2R	11895173...93,+	2	1	1			1		2	5	1	14
mir-31a	UGGCAAGAUGUCGGCAUAGCUGA	22-23	2R	12847576...98,–	1						1		1		2
mir-6-3§*	UAUCACAGUGGCUGUUCUUUUU	21-22	2R	14724442...63,–	9	1					1		1		10
mir-6-2§*	UAUCACAGUGGCUGUUCUUUUU	21-22	2R	14724584...615,–	9	1									10
mir-6-1§	UAUCACAGUGGCUGUUCUUUU	21-22	2R	14724732...53,–	9	1							5	1	10

(continued)

Table 2.2 microRNA profile during *Drosophila* development

Gene	Mature miRNA and miRNA* Sequence	Size Range (nt)	Chr		Position, Orientation	Embryo (hr) 0–2	2–4	4–6	6–12	12–24	Larva 1	1 + 2	3	A	T	Total
mir-5	AAAGGAACGAUCGUUGUGAUAUG	22–23	2R	|	14724920...42, –	3	5	1								9
mir-4[c]	AUAAAGCUAGACAACCAUUGAA	21–22	2R	|	14725020...41, –	5	2									7
	*CUUUGGUCGUCCAGCCUUAGGUGA	24–25	2R	|	14725053...76, –		2									2
mir-286	UGACUAGACCGAACACUCGUGCU	22–24	2R	|	14725155...77, –	1	7	3	2					1		14
mir-3[c]	UCACUGGGCAAAGUGUGUCUCA	18–22	2R	|	14725320...41, –	7	2									9
mir-309	GCACUGGGUAAAGUUUGUCCUA	22	2R	|	14725430...51, –		2									2
mir-310[a]	UAUUGCACACUUCCCGGCCUUU	22	2R	|	15647477...90, –											
mir-311[a]	UAUUGCACAUUCACCGGCCUGA	22–23	2R	|	15647800...21, –	1	1									2
mir-312[a]	UAUUGCACUUGAGACGGCCUGA	22–23	2R	|	15647770...91, –	2	1									3
mir-313[a]	UAUUGCACUUUUCACAGCCCGA	22	2R	|	15647905...27, –											
mir-7[a]	UGGAAGACUAGUGAUUUUGUUGUU	21–24	2R		15660793...816, +	3	1	1								5
bantam	UGAGAUCAUUUUGAAAGCUGAUU	20–23	3L		622988...010, +			2			1	2		3		8
mir-276b	UAGGAACUUAAUACCGUGCUCU	22	3L		10277375...96, +									1		1
mir-276a	UAGGAACUUCAUACCGUGCUCU	22	3L		10322810...31, +				2					6		8
mir-314	UAUUCGAGCCAAUAAGUUCGG	21	3L		11730620...40, +	1								2	1	4
mir-285	UAGCACCAUUCGAAAUCAGUGC	22	3L		11903655...76, –									1		1
mir-263b	CUUGGCACUGGGAGAAUUCAC	21	3L		15792330...50, –									1		1
mir-315	UUUUGAUUGUUGCUCAGAAAGC	22	3L		18809861...82, +					1						1
mir-9c[a]	UCUUUGGUUAUCUAGCUGUAUGA	21–23	3L		19515007...116, +	5	1	2	2			1			1	12
mir-316	UGUCUUUUUCCGCUUACUGGCG	20–22	3L		21366044...65, –								1		1	2
mir-10[a]	*CAAAUUCGGUUCUAGAGAGGUUU	23	3R		2635235...57, –							1				1
	ACCCUGUAGAUCCGAAUUUGU	21	3R		2635277...97, –	1										1
mir-317	UGAACACAGCUGGUGGUAUCCAGU	20–24	3R	|	5916921...44, +									5	3	8
mir-277	UAAAUGCACUAUCUGGUACGACA	21–23	3R	|	5925920...42, +									13		13
mir-34	UGGCAGUGUGGUUAGCUGGUUGUG	17–25	3R	|	5926692...715, +									6	7	13
mir-318	UCACUGGGCUUUGUUUAUCUCA	22	3R		6234084...105, +									3		3
mir-13b-1§	UAUCACAGCCAUUUUGACGAGU	17–24	3R	|	11243152...73, –	3					1					4
mir-13a	UAUCACAGCCAUUUUGAUGAGU	22	3R	|	11243287...308, –											
	*CUCCUCAAAGGGUUGUGAAAUG	22	3R	|	11243328...49, –		1									1
mir-2c[a]	UAUCACAGCCAGCUUUGAUGGGC	22	3R	|	11243500...23, –											
mir-iab-4-5p	ACGUAUACUGAAUGUAUCCUGA	22	3R		12662018...39, +		1									1
mir-iab-4-3p	CGGUAUACCUUCAGUAUACGUAAC	24	3R		12662052...75, +			1								1
mir-11	AUCACAGUCUGAGUUCUUGC	20–23	3R		17439190...209, –	6	1									7
mir-92a	CAUUGCACUUGUCCCGGCCUAU	22–25	3R	|	21461947...68, +	2	1									3
mir-92b	AAUUGCACUAGUCCCGGCCUGC	22	3R	|	21466514...65, +	1										1
mir-279	UGACUAGAUCCACACUCAUUAA	22	3R		25030739...60, +	1										1

Table 2.2 (continued)

◀ **Table 2.2 microRNA profile during *Drosophila* development**
Clusters of miRNAs are represented by vertical bars between the columns that indicate the chromosome location. An asterisk is used to denote small RNAs that are derived from the strand opposite to the miRNA strand within the fold-back precursor. For the iab-4 derived miRNAs, 5p and 3p indicate 5' and 3' location within the conserved fold-back sequence of the iab-4 transcript.
§ More copies of this miRNA are found in the genome and the clone numbers indicated cannot be assigned to a unique locus.
P Predicted miRNA based on phylogenetic conservation or vicinity to other clustered miRNAs.
n Expression of miRNA was also confirmed by Northern blotting.

2.2.6   Discovery of Repeat Associated silencing RNAs - rasiRNAs

A surprisingly large number of small RNAs (178 clones) mapped to repetitive sequence elements of the *D. melanogaster* genome. The relative ratio of miRNAs to rasiRNAs varies as a function of development. In the early embryo there are equal proportions of miRNAs to rasiRNAs, but the fraction of rasiRNAs drops significantly during the transition to adults, in which only about 10% of the small regulatory RNAs are represented as rasiRNAs. The content of rasiRNAs in microdissected testes is comparable to that of early embryos. Together, these findings suggest that the taming of transposable elements and the establishment of chromatin structure may be initiated in germline tissue and during early embryonic development in an RNA-dependent manner.

The rasiRNAs contain sequences from all known forms of repetitive sequence elements, such as retrotransposons, DNA transposons, satellite, and microsatellite DNA sequences, complex repeats such as the *Su(Ste)* locus, as well as vaguely characterized repetitive sequence motifs. The transposable elements are categorized by their mode of transposition and fall into two major subgroups. The class I elements transpose via an RNA intermediate, while the class II elements transpose through a DNA intermediate. Class I elements are again subdivided into long-terminal repeat (LTR) retrotransposons and non-LTR transposons, also referred to as long-interspersed nuclear elements (LINE) or poly(A)-type retrotransposons. Class II elements are characterized by a terminal-inverted repeat (TIR). We have cloned small RNAs from 38 different transposable elements corresponding to 40% of all known transposable elements in *D. melanogaster* (Kaminker, Bergman et al. 2002). The most frequently cloned rasiRNAs were derived from the LTR transposon *roo*, which is also the most abundant transposable element in the euchromatic portion of the *D. melanogaster* genome. This indicates a role for RNAi in controlling the mobility of transposable elements in *D. melanogaster*.

The production of small sense and antisense RNAs from the repetitive *Su(Ste)* locus sequences was shown previously (Aravin, Naumova et al. 2001), indicating a role for dsRNA in the homology-dependent regulation of *Stellate*. In agreement with this, we cloned several siRNAs in sense and antisense orientation from the *Su(Ste)* locus in testes tissue. The identification of rasiRNAs from other not yet annotated repetitive sequences suggests that systems similar to that of *Ste/Su(Ste)* might exist in *D. melanogaster*. Therefore, small RNA cloning is a viable strategy for identification of such cases.

Some repeat sequences such as HeT-A, TART, and subterminal minisatellites are restricted to telomeres and required for telomere maintenance. These elements are absent from euchromatic regions (Kaminker, Bergman et al. 2002) and therefore link small RNAs to regulation of heterochromatin. Small RNAs of the GAGAA microsatellite family were cloned from testes and originated probably from transcripts of the fully heterochromatic Y chromosome (Lohe, Hilliker et al. 1993). Another group of 10 rasiRNAs could be assigned to a 200 kb region from chromosome 2R (band 42AB), which is composed of numerous different types of transposable elements, many of which are damaged and have diverged from canonical sequences. There are no predicted genes in this region, but several lethal mutations have been described that map to this region (Bender, Kooh et al. 1993). The reasons for this lethality are unclear, but it is conceivable that this region is an important source for small RNAs involved in some aspect of heterochromatin regulation.

Our set of 178 rasiRNAs represents the largest collection of repeat-associated small RNAs (rasiRNAs) to date. It provides insights into the complexity of RNA-guided regulation of heterochromatin. By analogy to the plant world, it is presumed that this longer class of siRNAs is also mediating transcriptional regulation. *Ago3*, *aub*, and *piwi* are predominantly expressed in germ cells and the early embryo (Williams and Rubin 2002). The increased abundance of rasiRNAs in embryo and testes strongly suggests that these Argonaute members are specific binding partners for rasiRNAs. This discovery was borne out in work done some years after the paper was written (Aravin, Hannon et al. 2007; Brennecke, Malone et al. 2008)

2.3   microRNA targets in *Drosophila melanogaster*

2.3.1   Background to target prediction

The work presented in this section was the first microRNA target prediction study published and went online in Genome Biology October 2003. Stark et al, also published a paper on microRNA target prediction in *D. melanogaster* in the same month (Stark, Brennecke et al. 2003). We provided specific predictions about microRNAs that might regulate individual genes, and offered, for the first time, a sense of the scope and specificity of microRNA regulation. A key finding from this work was that the microRNA regulation of mRNA function is combinatorial. I present the work as it was developed at the time, built on the then *limited* knowledge of microRNA:mRNA regulation. Many of the predictions from this work have since been validated.

At the time of this study, several hundred different microRNAs had been identified from various organisms and their sequences were archived and accessible at the RFAM miRNA registry website. This database then contained 21 miRNAs from *Arabidopsis thaliana*, 48 from *C. briggsae*, 106 from *C. elegans*, 73 from *D. melanogaster*, 122 from *Mus muscularis*, and 130 from *Homo sapiens*. The two founding members of the microRNA family, lin-4 and let-7 in *C. elegans*, were discovered genetically with the what will be Nobel prize work showed them with central role as key regulators of developmental timing through cell fate decisions (Lee, Feinbaum et al. 1993; Wightman, Ha et al. 1993). These miRNA genes are also conserved in other animals and mammals, and the miRNA binding sites on homologous genes were also conserved. A function had also been identified for the bantam miRNA; it was found to regulate cell proliferation and cell death by targeting the apoptosis gene hid (wrinkled) (Brennecke, Hipfner et al. 2003). The *D. melanogaster* miR-14 had been implicated in fat metabolism and stress resistance as well as cell death (Xu, Vernooy et al. 2003), however the precise target genes of this miRNA had not been identified.

Therefore, in 2003, hundreds of microRNAs had been discovered but the scope of their regulation was unknown. The discovery of ~ 70 microRNAs in *Drosophila* melanogaster, Chapter Two, naturally raised the question of what they did. Positive selection of the microRNAs over the 40 million years between *D. melanogaster* and *D. pseudoobscura*, meant that the functions were likely to be conserved too, and important for the organism. We know from the few known microRNA-target relationships, that microRNAs target mRNAs through imperfect base-pairing with the 3'UTR. It is the nature of the 'imperfectness' that provides the computational challenge to predict the matching functional microRNA–mRNA pairs. When we started this work only five microRNA targets were known and only one of these in *Drosophila* (*bantam* and *hid*). The characteristics of the base pairing, such as internal bulges and mismatches and wobbles, made it important to design a user-flexible approach to the problem. The ability to add new rules, which were inevitable come from individual experiments, made it necessary that the method also be modular. This rest of this chapter describes the algorithm, miRanda, for microRNA target prediction and the functional predictions for microRNAs in the fly.

2.3.2   Architecture of microRNA target sites

Only five known microRNA:target mRNA interactions were known when the algorithm described here was developed: (i) lin-4:lin-14 (in worm)(Lee, Feinbaum et al. 1993) (ii) lin-4:lin-28 (Moss and Tang 2003) (iii) let-7:lin-41 (in worm) (Reinhart, Slack et al. 2000) (iv) bantam:hid (in fly) (Brennecke, Hipfner et al. 2003) and (v) let-7:hbl-1 in worm (Abrahante, Daul et al. 2003). These five interactions contained around a total of 15 proposed target sites. The pairing of the nucleotides in the microRNA and the target mRNA to form a (stable) duplex.

A word of caution about the interpretation of these target pairs; Although the target pairs are experimentally validated, the details of the binding site duplex were not determined, so that the duplexes shown are predictions. To varying degrees these experiments show that the particular sites or combinations of them are needed, but they do not show that the actual binding configuration is the one which is proposed in the RNA duplex thus calculated in silico. Indeed, there is a large body of work showing that RNA is able to tolerate more mismatches in a helical formation without the expected compensatory loss of binding energy (as compared to DNA). (refs).

In the five known interactions bewteen microRNAs and their targets, referenced above, the RNA:RNA duplexes have several features apparent on manual inspection (see refs Lee, Feinbaum et al. 1993, Moss and Tang 2003 Reinhart, Slack et al. 2000, Brennecke, Hipfner et al. 2003 and Abrahante, Daul et al. 2003).

(i) Mismatches or "loop-outs" in middle portion. Most of the first known targets sites showed little or no matching for positions 8-12 of the microRNA and yet the loop-out regions are conserved; lin-4 within worms, bantam within flies and let-7 across all metazoa. The mismatches from 8-13 were in contrast to siRNA functional requirements; exogenous siRNAs used a full complementary match to their target UTR including the central portion. In particular, a minimal requirement for the siRNA to direct endonucleolytic cleavage of the target mRNA opposite 10-11 of the siRNA was a perfect match at this position,

(ii) 5' end domination of match. The proposed target sites suggested a stronger requirement for matching positions m1-m9 than the remainder of the microRNA. Lee et al identify seven matches lin-4 to lin-14 with the five prime match section being the most similar, and the most conserved (in *C. briggsiae*) between the sites (Figure 2.1). Four of the seven proposed duplexes have an 'insertion' at m6 and one at m2 both of which would cause a bulging of the nucleotide, out of the duplex. (Note that an 802 nt deletion which contains all 7 sites only partially reduces the Lin-14 gradient and similarly reporter genes with all, 7 sites showed only partial temporal gradient compared to whole UTR (Wightman 1993).

(iii) Possible requirement for an insertion in 5' region. Four of the lin-4 sites in lin-14 have a bulging C at position 6 of the microRNA. When the bulging C at 6 is mutated to a U the mutated U may now base pair with A and disrupts the overall base paring in the remainder of the duplex. However, since other microRNA: mRNA pairs do not require this mismatch, we do not make it a requirement of the algorithm.

(iv) G:U wobbles. The C to U transition mutation at position 5 in lin-4 mutant-did not retain full physiological function (Ambros 1993), yet there are G:Us in a number of these sites. Allowing the the G:U wobble pairing may be context dependent. In the position 5 mutant the G:U is adjacent to the bulging C at position 6 causing disruption to the A helical form and this amy explain the lack of tolerance.

The sites showed a weaker duplex requirement in positions 16–24 (the '3' end) than 1-9 (the 5' end) and none in postions 10–15.

2.3.3 Number and location of binding sites in the target mRNA

microRNA target genes contain more than one site:
These four initial studies did not reveal obvious requirements for a specific

number of sites or positions of binding sites; they vary between one and five potential sites and the distances between the sites show no obvious pattern, However, the experiments conducted revealed some constraints and tendencies.

Additivity: The number of sites in the UTR affects the amount of down-regulation.. Wightman et al.,showed an element bearing 3 of the 7 conserved lin-4 sites in the 3' UTR of a reporter showed a partial, not full, temporal gradient.

Cooperativity: These early studies proposed cooperativity to explain the shape of the temporal gradient. In particular, synergy between the seven lin-4 sites could some of the sharp down-regulation in Lin-14 protein level at specific lin-4 concentrations.

All the predicted microRNA sites were in the 3'UTR of their respective target genes. Reconstruction of the effect of the target sites with the sites placed in the three prime UTR of the construct, also supported that this is the 'natural' place for animal microRNA target sites. Attempts to downregulate target genes including reporters by placing sites in the 5'UTR were not successful. There are no reports of animal sites tested in coding regions, and this may reflect the fact that researches have tried and failed, or not tried.

There is no consensus on the positions of the binding sites relative to each other or relative to the length of the transcript. However, the experiments of Doench et al, placed the sites at 27 nucleotides apart (measured from 5' to 5' of the sites), which may have turned out to be fortuitously optimal for synergy.. Many later studies have supported the observation that proximity of sites causes synergy of effect, in particular a 10-40 nt spacer between microRNA sites (Saetrom, Heale et al. 2007).

2.3.4   siRNA studies demonstrate cooperativity of target sites

Since siRNAs can behave like microRNAs (Doench and Sharp 2004), we could use the information from siRNA design to learn about microRNA targeting. The connection between 'siRNA effect' and 'microRNA effect', and its implications for the design of si- and sh-RNA specificity has far reaching implications for both RNAi functional studies and for the potential use of si- and sh-RNAs as therapeutics. I will return to the implications of the results of these studies for microRNA function in Chapter 3) Doench et al in 2003 provided evidence Increasing the number of bulged siRNA/miRNA-binding sites in the 3'UTR of a reporter gene increases the degree of repression and this may be more than additive. Zero, two, four and six sites imperfect (and perfect) complementarity sites were cloned into the 3' UTR of into Renilla reniformis luciferase and transfected into HeLa cells. The results showed increasing repression for increasing numbers of sites for both the partially complimentary and fully complimentary- siRNA-like- constructs. However, for the partial complimentary sites- the microRNA-like sites- four sites had twofold more repression per site than two sites, Figure 2.2 A and B whereas the siRNA-like constructs produced increasing repression but not increasing repression per site, Figure 2.2 C and D.

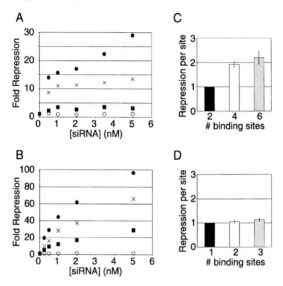

Figure 2.2 Cooperativity of miRNA binding sites

2.3.5   Conservation of microRNAs and implications for target sites

The striking conservation of the full-length let-7 was noted by Pasquinelli et al., in 2000 (Pasquinelli, Reinhart et al. 2000)and this was followed by many other examples in the cloning of these small RNAs in October 2001. Since miRNA-containing species have been separated by hundreds of millions of years of evolution, it is striking that many 22 nucleotide miRNAs do not exhibit stronger sequence divergence. (Although lin-4 is not fully conserved in flies or vertebrates, it does have a vertebrate homolog, miR-125b, which is identical positions 1–11, 14–16 and 19–23.)

About 20% of the miRNAs identified in invertebrates are conserved in mammals, and about 10% of mammalian microRNAs are conserved in invertebrates (in 2004). This suggests that the regulatory function of these genes may be conserved across this large phylogenetic distance. More recent studies and our own using 2007 RFAM (#ref) found ~ 30 miRNA 'families' are conserved across the bilateria and were therefore present in urbilateria, the last common ancestor of all living bilaterians (Wheeler, Heimberg et al. 2009),. Members of the same family are defined by identical sequence from positions 2–8 on the basis that these microRNAs have overlapping targets and functions (Bartel 2004, however its likely that this definition of a 'family' is glossing over more major differences between same-seed groups than is currently understood (Leaman, Chen et al. 2005; Jackson, Burchard et al. 2006).

This absence of sequence-evolution in many micoRNAs supports a model where each of these conserved microRNAs may have tens, if not hundreds, of target sites and hence that evolution by compensatory base-pair changes has become extremely unlikely. It follows then, that a microRNA may regulate few or many genes depending on its apparent birth date. This leads to the prediction that there will be a distribution of targets per microRNA which reflects the extent of conservation of the microRNA itself.

The seven initial studies described above in this chapter identified some 15 target sites with varying degrees of conservation, Table 2.2. The question is, are most sites which are in one species conserved in the orthologous mRNA of another, somewhat closely related species?

80% of the sites between *C. elegans* and *C. briggsiae*, as well as *D. melanogaster* and *D. pseudoobscura* in these initial studies were conserved with ~ 70% conservation i.e. around 4-5 nucleotide difference. The differences between the sites across species are mostly insertions/deletions at positions 10–15 in the microRNA. The sites are mostly conserved in positions 1-9 and 16-22. Significantly, these conserved positions were coincident with positions which were more likely to be base paired (with the notable exception of the bulging C in the lin-4 example, in reference Lee, Feinbaum et al. 1993.

Strong additional support of the significance of the conservation is functional conservation of the interaction between microRNA and it's target mRNA(s).

In the prescient work by Lee et al., they examined the lin-4:lin14 interaction in 3 other *Caenorhabditis* species and the rescue of the *C. elegans* lin-4 null allele suggests that the interaction is conserved. Similarly, Reinhart et al, show that a 2.3-kb genomic DNA fragment from *C. briggsae* complemented let-7 in *C. elegans*, showing that let-7 function is conserved between the two species.

2.3.6 Design of the target prediction algorithm, miRanda

Our goal was to develop a target prediction method, which had the following features:

(i) Given an mRNA sequence, find matching sites (target sites) for every microRNA. Thus, for every microRNA and mRNA the algorithm should find a ranked list of maximally complementary sites. 'Maximally' refers to optimizing some kind of complementarity score that reflects the apparent principle of recognition as seen in the so far experimentally observed cases.

(ii) Predict known targets with high reliability.

(iii) Sensitive enough to give a reliable indication of the scope/extent of microRNA regulation.

(iv) Specific enough to avoid massive over-prediction

(v) Have adjustable parameters that have biological meaning. It should be modular enough for users to tweak individual parts of the pipeline without compromising the output; it should be easily extended and integrated with other biological data e.g. expression levels.

Reliable identification of microRNA targets is qualitatively different from standard sequence similarity analysis and requires new methods. In traditional sequence analysis, one tries to assess the likelihood of a hypothesis, for example, whether similarity between two sequences is due to common ancestry and continuity of functional constraints or a chance occurrence. In microRNA target prediction, however, the aim is to assess the likelihood of a physical interaction between two molecular species: a mature microRNA, and a full length and translation-competent mRNA. Detecting this physical interaction can gain from a two-pronged approach; one which considers only the inverse sequence match, and the other which considers the thermodynamics of binding. The goal of the inverse sequence match is to identify a stretch of nucleotides that is complementary to a potential partner (as opposed to identifying nucleotide sequences that give the same or similar nucleic acid or amino acid sequence, which is the goal of many sequence comparison algorithms). The goal of the physical binding search is to identify thermodynamically favorable duplexes.

The challenges to achieve the goals in (i)-(v) are many and include: defining the score to be optimized, defining the degrees of freedom, i.e., the configuration space over which optimization occurs (search space), the optimization algorithm and the balance of sensitivity versus specificity.

In approach developed here, the problem is divided into modular components: target site detection and scoring, duplex energetics, conservation, overall rank/score (for each microRNA and for each gene), and finally validation statistics.

The target site identification is a search in sequence space and this is reminiscent of the well studied sequence alignment problem. Here, the same type of dynamic programming method used for most sequence alignment problems is adapted to the problem at hand.

The dynamic programming method for sequence alignment is a particular induction method, developed over the years to deal with mismatches, with deletions/insertions of single symbols and, eventually, with insertions/deletions of longer segments (Needleman and Wunsch 1970; Smith, Waterman et al. 1981). The similarity value S between two match candidates at one position reflects either the minimal number of mutational events to convert one sequence symbol into another or, more generally, the likelihood of accepted point mutations based on statistical evaluation of known correctly matched sequences or on physico-chemical similarities of the polymer units (e.g., for amino acids). The total alignment score for two sequences is assumed to be a simple sum of S values over the entire alignment trace. In the local Smith-Waterman algorithm variant (Smith and Waterman 1981) used here, the algorithm starts at the beginning of the sequences and builds up a 'best partial alignment' matrix A in N by M decision steps, for sequences of length N and M. The key repeated algorithmic step is the extension of the best partial alignment up to position (i,j) in the two sequences by a local decision that chooses between a sequence match at $(i+1,j+1)$, a deletion at i or a deletion at j (assuming reasonable values for the deletion events). The deletions can be of finite length k, using a gap-open penalty plus linear gap penalty proportional to k-1. In addition to additivity of the similarity/deletion scores, the algorithm assumes that the local decision at (i,j) is independent of any previous decision. With that assumption, the complexity of the algorithm is of order N * M. After completion of the A matrix, a back tracing procedure from the matrix element with the largest total alignment score is used to generate the explicit trace of the aligned sequence.

In adapting the Smith-Waterman algorithm to the miRNA target site prediction problem, an additional device use here is the introduction of a position-specific weight $w(i)$ phased on one of the sequences such that the similarity score S at position i is replaced by $w(i) * S$. In practice, this algorithmic extension works well for the purpose of taking the match at certain positions more seriously than at other positions, a key aspect of the physico-chemistry of small RNA matches to messenger RNAs in the context for the RISC complex.

The miRanda algorithm is an adapation of the Smith-Waterman algorithm as outlined above. However, instead of building alignments based on matching nucleotides (A-A or U-U, for example), it scores based on the complementarity of nucleotides (A=U or G=C). The scoring matrix used for this analysis also allows G=U 'wobble' pairs, which are important for the accurate detection of RNA:RNA duplexes. Complementarity parameters at individual alignment positions are: +5 for G≡C, +5 for A=U, +2 for G=U and -3 for all other nucleotide pairs, and these can be easily set by the user, Figure 2.3. Other non-canonical base-pairings are possible in A helical RNA duplexes, however, these seem to be contingent on the specific identity of neighboring base pairs and may not allow the conformational flexibility which this duplex formation may require. When miRanda was developed, little was known about any of the conditions which may affect this: e.g. the binding kinetics, binding, the structure of the formed duplex, or the holding conformation of the single-stranded microRNA in the Argonaute protein (as discussed in Chapter 1). Therefore, we chose not to include them in the match search for the time being.

The algorithm uses affine penalties (linear in the length of a gap after an initial opening penalty) for gap-opening (-8) and gap-extension (-2). In addition, following observation of known target sites, complementarity scores (positive and negative values) at the first eleven positions were multiplied by a scaling factor $w(i)$, for $i=2$-11, (here set at 2.0), so as to reflect the observed 5'-3' asymmetry, Figure 2.3. As more experimentally validated targets were published after the initial publication of miRanda, the default parameters were refined to reflect this data. We investigated a range of weightings ($w=1, ...,6$) and assessed the effects on the rank and score of known target sites, and the randomization statistics. As a result, we changed the default weighting parameter to $w=4$ and weighted only nucleotides in microRNA positions $i=2$-9.

With these parameters, the dynamic programming algorithm optimizes the complementarity score between a miRNA sequence and an mRNA sequence

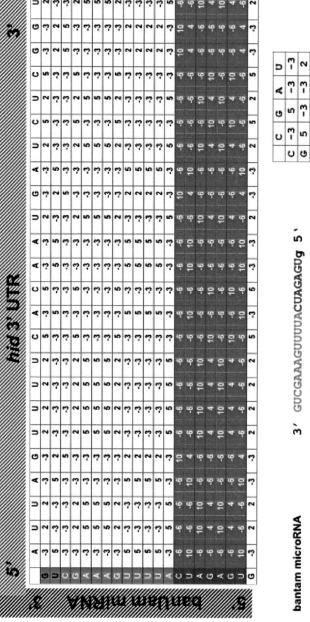

bantam microRNA

hid 3'UTR

3' GUCGAAGUUUACUAGAGUg 5'
    : ||:||||| ||||||||:
5' UAGUUUCACAAUGAUCUCGg 3'

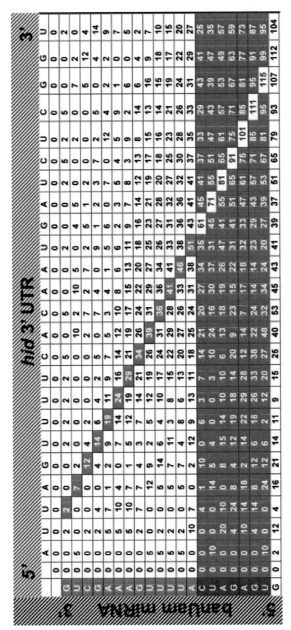

Figure 2.3 Fine tuning the weighting.

(typically a 3' UTR), summed over all aligned positions, and finds all non-overlapping hybridization alignments in decreasing order of complementarity score down to some cutoff value (default value 80). The detection of sub-optimal alignments follows heuristics previously used in sequence alignments (Comet and Henry). In the first instantiation of miRanda, the following four empirical rules were applied: (i) no mismatches at i=2-4 (ii) fewer than five mismatches between i=3 and i=12 (iii) at least one mismatch between i=9 and i=n-5) (where n is the length of the microRNA) (iv) fewer than two mismatches in the last five positions of the alignment.

After manual inspection of the results of these rules on the target sites and experimentally validated targets published after 2003, we refined our default parameters and simplified the empirical rules.

To significantly increase the speed of miRanda runs, in calculating the optimal alignment score at positions i, j in the alignment scoring matrix, the gap-elongation parameter was used only if the extension to i, j of a given stretch of gaps ending at positions i–1, j or j–1, i (but not of stretches of gaps ending at i–k, j or j, i–k for k > 1) resulted in a higher score than the addition of a nucleotide–nucleotide match at positions i, j. Removal of this restriction with the availability of more computing power would result in a moderate increase in average loop length, but the advantages of this would probably be superseded by overall refinement of target prediction rules. Importantly, complementarity scores at the first eleven positions, counting from the miRNA 5' end, were multiplied by a scaling factor of w(i)=2.0, so as to approximately reflect the experimentally observed 5'–3' asymmetry; for example, G:C and A:T base pairs contributed +10 to the match score in these positions. The value of the scaling factor at each position is an adjustable parameter subject to optimization as more experimental information becomes available.

2.3.7   Assumptions with the miRanda algorithm

(i) That there is position independence – i.e. the sum is the linear addition of the weights which are fixed on one position. This is a standard assumption of sequence alignment methods and we decided to return to this when more about interaction was known.

(ii) Changes to the relative weighting does not violate the algorithm. Although changing the relative weighting is not traditionally used in sequence alignment, it is consistent with the basic assumptions in the dynamic programming algorithm. This is because the algorithm assumes additivity of score, the decision n>n+1 does not depend on previous best solution at n, as in (i).

(iii) Protein interactions with mRNAs do not affect the likelihood of binding. This is unlikely to be true but we cannot know more detail until more is known mechanistically.

2.3.8   Free energy of RNA duplex

Given the drawbacks of sequence complementarity alone, we decided to use the physical interaction as part of the target identification pipeline. To detect a potential binding of the local mRNA to the microRNA, we initially proposed using a 'net' binding energy, where $\Delta Gnet = \Delta Gduplex - \Delta GmRNA\_nts$

i.e. the difference between free energy of the interaction itself and the free energy of the nucleotides from the target mRNA in the UTR.

We examined the local RNA secondary structure of known targets to see whether there were detectable requirements e.g. accessibility of the binding nucleotides. Our prediction was that known target sites might be in loop regions rather than predicted duplex, regions of low structure stability, or just homogenous between sites in some recognizable way.

The results of our investigations were inconclusive. The predicted secondary structure sequence produced different secondary structures of target sites with differing lengths of input sequence from the UTR and the target sites showed no uniformity conformation. We concluded that we would return to this when more target sites had been discovered and we had made a first set of predictions to test.

However, we did retain the score for the free energy of the duplex and for each match, the free energy (ΔG) of optimal strand-strand interaction between miRNA and UTR is calculated using the Vienna package. In order to estimate the thermodynamic properties of a predicted duplex, the algorithm uses folding routines from the Vienna 1.3 RNA secondary structure programming library (RNAlib) (Wuchty, Fontana et al. 1999). The expanded thermodynamic parameters used are more computationally intensive than the initial scan, but allow potential hybridization sites to be scored according their respective folding energies. The miRNA sequence and 3' UTR sequence from a hybridization alignment are joined into a single sequence with an eight base sequence linker containing artificial 'X' bases that cannot base pair. This strand-linker-strand configuration assumes the phase space entropy of strand-strand association is constant for all miRNA-target matches (Wuchty, Fontana et al. 1999). The minimum energy of this structure with the last matching base pair (from initial sequence alignment) constrained is then calculated using RNAlib.

2.3.9   Conservation rules

We built the miRanda pipeline such that varying degrees of conservation stringency could be tuned up or down, Figure 2.4.

However there are at least three somewhat independent aspects to conservation,

(i) Existence of a target site above threshold in a different species; issues include phylogenetic distance of comparison species

(ii) Position of target site in aligned UTRs; issues – is this 'reaslistic'? how well do current algorithms align UTRs?

(iii) Sequence conservation of site, i.e. pattern of binding conservation. For instance, we could include that the conservation of the site has the same 'shape'. Issues, 'patterns' may be very susceptible to single nucleotide changes. The question of how closely related the organisms should be was solved heuristically, and is discussed in full in below, with respect to insect genomes.

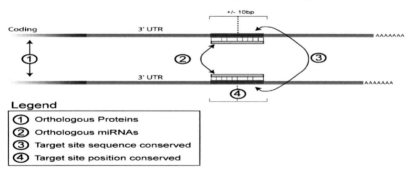

Figure 2.4 Conservation rules

2.3.10   Benchmarking and validation

Validation of the method would be difficult since we do not know the null model, nor precedent for this kind of sequence analysis. Statistical validation makes the assumption of what to compare it against, and may make the assumption that the observations should be rare.

The initial validation of this target prediction method was that it correctly predicted known target sites. We acknowledged at the time that this was an unsatisfactory validation method, since the algorithm itself was developed using the same target sites as a guide. Independent validation and testing of our specific hypotheses came later.

We initially showed that the method correctly identifies nine of the ten currently characterized target genes (Table 2.2) for the miRNAs lin-4 and let-7 in *C. elegans* and bantam in *D. melanogaster*. At the chosen ΔG threshold level, the details (position and base pairing as reported by others) of most target sites were largely reproduced, together with interesting alternative target sites on the known target genes. This comparison is only partially conclusive, as not all reported target sites on known target genes have been individually verified by experiment. The missed duplex between the lin-4 miRNA and its reported target gene (lin-14) (Table 2.2) contains an unusually long loop structure in the target sequence, which cannot easily be detected without adversely affecting the rate of false positive detection. The rankings obtained from our additive scoring scheme for these targets were also consistently high. For example, the two target genes of the let-7 miRNA (hbl-1 and lin-41) are detected as the number 1 and number 2 ranked genes hit respectively, from a scan against 10,000 *C. elegans* 3' UTR sequences.

Progress on the identification of animal miRNA targets had been slow and computationally difficult because animal miRNAs are only partially complementary to their mRNA targets. Developing algorithms to predict such target sites is therefore difficult In contrast, plant miRNA targets were easier to identify because of near-perfect complementarity to their target sequence (Rhoades, Reinhart et al. 2002). Many of these plant mRNA targets encode transcription factors that regulate morphogenesis (Llave, Xie et al. 2002; Palatnik, Allen et al. 2003; Tang, Reinhart et al. 2003) (Xie, Johansen et al. 2004). As a consequence of near-perfect complementarity, plant miRNAs predominantly act as small interfering RNAs (siRNAs) guiding destruction of their mRNA target, though some have also been found to behave like animal miRNAs (Chen 2004). Plant miRNA target sites identified up to this point were predominantly found within the protein-coding segment of the target mRNAs, while animal miRNAs appeared to primarily target the 3' untranslated region (UTR). About 10% of the miRNAs identified in invertebrates are also conserved in mammals, indicating that the regulatory function of these genes is likely to be conserved cross-species. Since miRNA-containing species have been separated by hundreds of millions of years of evolution, it is striking that many 22 nucleotide miRNAs do not exhibit stronger sequence divergence. This absence of sequence-evolution in many miRNAs suggests that these miRNAs have more than one target site and that evolution by compensatory base-pair changes has become extremely unlikely. Therefore, a miRNA may regulate few or many genes depending on its apparent birth date. It is also conceivable that additional evolutionary constraints, such as the presence of certain protein-binding sites within the miRNA-targeted mRNAs, are conferring specificity to these small RNA regulated processes.

Given imperfect rules for sequence pairing and energy estimation, the conservation of predicted miRNA-target pairs in closely-related species is an important additional criterion for this analysis. Given the surprisingly high level of sequence conservation of miRNAs across phyla (Griffiths-Jones 2004), we make the assumption that the set of miRNAs in *D. melanogaster* is largely shared with *D. pseudoobscura* and A. gambiae for the purposes of our predictions.

2.4   microRNA target prediction in *D. melanogaster* using miRanda

2.4.1   Collecting sequences and running miRanda

An initial set of *D. melanogaster* miRNA sequences was built using the RFAM miRNA database from 2003 (Griffiths-Jones 2004). Mature miRNA sequences were placed in a FASTA formatted sequence file. In total, the final file contained 73 unique miRNA sequences. Sequences for *D. melanogaster* 3' UTRs were obtained from the Berkeley *Drosophila* Genome Project (BDGP). In total, 3' UTR sequences were available for 14,287 transcripts, representing 9,805 individual *D. melanogaster* genes. A corresponding set of *D. pseudoobscura* 3' UTR sequences was then built from the March 2003 first freeze of the *D. pseudoobscura* genome project at Baylor College of Medicine. Each *D. melanogaster* 3' UTR was mapped to *D. pseudoobscura* contigs by searching both the actual *D. melanogaster* 3' UTR sequence, using NCBI BLASTn (Waterman and Eggert 1987), and the peptide sequence of each gene, using NCBI tBLASTn, against

D. pseudoobscura contigs (Waterman and Eggert 1987). Results from these two scans were then used to identify candidate 2,000 bp regions of *D. pseudoobscura* contigs, within which we believe an orthologous *D. pseudoobscura* 3' UTR is present. The AVID (Bray, Dubchak et al. 2003) alignment tool was used to align the real *D. melanogaster* 3' UTR and a candidate *D. pseudoobscura* region. Finally, this alignment was used to trim each candidate region, leaving the predicted *D. pseudoobscura* 3' UTR. In total 12,416 transcripts and 8,282 genes from *D. pseudoobscura* were mapped to orthologous *D. melanogaster* UTRs in this fashion.

The Ensembl database Application Programming Interface was used to construct *A. gambiae* predicted 3' UTRs by taking 2,000 bp downstream from the last exon of each transcript. Orthology mappings between *A. gambiae* and *D. melanogaster* UTRs were then obtained by searching all Ensembl *A. gambiae* peptides against all *D. melanogaster* peptides using BLASTp. In total 9,823 *A. gambiae* genes were mapped to *D. melanogaster* genes.

All miRNA sequences were scanned against the 3' UTR datasets of *D. melanogaster*, *D. pseudoobscura* and *A. gambiae*. The thresholds used for hit detection are: initial Smith-Waterman hybridization alignments must have $S \geq 80$, and the minimum energy of the duplex structure $\Delta G \leq -14$ kcal/mol. Each hit between a miRNA and a UTR sequence is then scored according to the total energy and total score of all hits between those two sequences. Hits are deemed to be conserved in *D. pseudoobscura* or *A. gambiae* if a target site equivalent to that detected in a *D. melanogaster* UTR can be found in the orthologous *D pseudoobscura* or *A. gambiae* UTR at the same position in the UTR alignments. Our definition of equivalence between target sites is that their sequences are more than 80% identical for *D. pseudoobscura* and 60% identical for *A. gambiae*. All results from the scan are then ranked and sorted according to total score of conserved target sites detected.The pipeline is shown in Figure 2.5. For each miRNA, the ten highest ranked genes are selected as its candidate target genes in this way. Multiple miRNAs binding the same site on a target gene are resolved using a greedy algorithm that assigns the highest scoring and lowest free energy miRNA target duplex to each potential site so that different miRNA target sites cannot overlap.

2.4.2   Validation

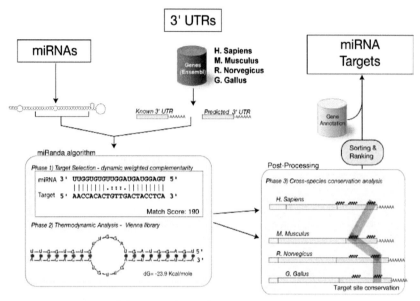

Figure 2.5 Overview of the miRanda pipeline
*Algorithm and analysis pipeline. Source data consisting of **(a)** miRNAs and **(b)** 3' UTRs are processed initially by **(c)** the miRanda algorithm, which searches for complementary matches between miRNAs and 3' UTRs using dynamic programming alignment (Phase 1) and thermodynamic calculation (Phase 2). **(d)** All results are then post-processed by first filtering out results not consistently conserved according to target sequence similarity with D. pseudoobscura and A. gambiae (Phase 3), then by sorting and ranking all remaining results. **(e)** Finally, all miRNA target gene predictions are annotated using data from FlyBase and stored for further analysis.*

The initial validation of this target prediction method was that it correctly predicted known target sites. We acknowledged at the time that this was an unsatisfactory validation method, since the algorithm itself was developed using the same target sites as a guide. Independent validation and testing of our specific hypotheses came later, see below.

We initially showed that the method correctly identifies nine of the ten currently characterized target genes (Table 2.2) for the miRNAs *lin-4* and *let-7* in *C. elegans* and *bantam* in *D. melanogaster*. At the chosen ΔG threshold level, the details (position and base pairing as reported by others) of most target sites were largely reproduced, together with interesting alternative target sites on the known target genes. This comparison is only partially conclusive, as not all reported target sites on known target genes have been individually verified by experiment. The missed duplex between the *lin-4* miRNA and its reported target gene (*lin-14*) (Table 2.2) contains an unusually long loop structure in the target sequence, which cannot easily be detected without adversely affecting the rate of false positive detection. The rankings obtained from our additive scoring scheme for these targets were also consistently high For example, the two target genes of the *let-7* miRNA (*hbl-1* and *lin-41*) are detected as the number 1 and number 2 ranked genes hit respectively, from a scan against 1,014 *C. elegans* 3' UTR sequences.

For the initial validation, 3' UTR sequences for *C. elegans* and *C. briggsae* were obtained, if possible, from UTRdb (Pesole, Liuni et al. 2002). If unavailable, UTR sequences were estimated by taking 2,000 bp of flanking nucleotide sequence downstream of the last exon of the gene in question using the Ensembl database (Hubbard, Barker et al. 2002).

Using intermediate thresholds (S: 80; ΔG: -14 kcal/mol), for each known miRNA and target gene pair (in either *C. elegans* or *D. melanogaster*), we list the number of known experimental target sites, the number of sites detected here, both raw and conserved in *C. briggsae* or *D. pseudoobscura*; and, the number and percentage of known sites that correspond to computationally detected conserved sites, with larger values indicating more successful (retrospective) prediction (†️ and 'N/A' indicate that no 3' UTR was available to scan against in *C. briggsae*, hence no conservation analysis was possible, results assume conservation). cel/ cbr: *C. elegans/C. briggsae*; dme/dps: *D. melanogaster/D. pseudoobscura*

To determine the likelihood that the targets identified were correct, we needed a set of control sequences to provide comparisons. Control sequences were constructed by assembling 100 sets of 73 miRNAs each generated by random shuffling of each *D. melanogaster* miRNA. Each of these sets of 73 randomized miRNAs was independently searched against all *D. melanogaster* and *D. pseudoobscura* 3' UTRs as in the reference experiment. Results and counts were then averaged over all 100 random sets, and were compared with the results of the actual miRNA scan, Table 2.3. For the functional analysis, GO classes for known *D. melanogaster* genes were obtained from FlyBase and conserved hits for the real and random miRNAs for each class are counted. The Z-scores are generated from the actual miRNA counts, averaged random miRNA counts and their standard deviations.

Organism	Target gene (3' UTR)	Number of experiment al sites	Number of predicted sites	Rank	Number of predicted sites with conservati on	Matches on conservati on	Matches in experiment al to predicted	Matches in experiment al to predicted (%)
cel/cbr	lin-14 (Abnormal cell-lineage protein 14)	7	1		0		0	0%
cel/cbr	lin-28	1	1	4/1,014	1		1	100%
cel/cbr	lin-41a lin41b	1	1	5/1,014	N/A		1+	100%+
cel/cbr	lin-14 (Abnormal cell-lineage protein 14)	2	6	9/1,014	2		2	100%
cel/cbr	lin-28	1	1	12/1,014	1		1	100%
cel/cbr	lin-41a lin41b	2	6	2/1,014	N/A		2+	100%+
cel/cbr	daf-12	3	10	7/1,014	1		1	33%
cel/cbr	hbl-1 (hunchback -related protein)	8	14	1/1,014	8		5	63%
dme/dps	hid (Head involution defective (wrinkled))	2	2	1/11,318	2		2	100%
dme/dps	CG10222	1	1	4/11,318	1		1	100%

Table 2.2 Validation of prediction method on experimentally verified miRNA targets

	Total hits	Total conserved hits	1 site	≥ 2 sites
73 D. melanogaster miRNAs (A)	6,864	589	556	33
73 Random miRNAs (B)	5,152	204	201	3
Standard deviation (100 experiments)	± 132	± 43	± 40	± 3
Ratio (A/B)		2.9	2.8	11.0
Estimated false positives (%)		35%	36%	9%

Detected conserved hits (especially those with multiple detected sites in the 3' UTR) are significantly over-represented (2.8× and 11× as many cases, on average, respectively) in analyses with actual miRNAs compared to randomly shuffled miRNAs. The thresholds used for this analysis were S: 100; ΔG: -19 kcal/mol; ID: 70%.

Table 2.3 Whole genome comparison of real versus randomized miRNAs against the complete genomes of *D. melanogaster* and *D. pseudoobscura*

2.4.3 Results: Scope of microRNA regulation

In our 2003 paper, we reported 535 potential target genes for the 73 known *D. melanogaster* miRNAs in decreasing order of match score for sites in detected 3' UTR targets. All of these targets passed the filters for free energy estimates, and conservation of target site sequence and position between *D. melanogaster* and *D. pseudoobscura*.

Of these predicted target genes in *D. melanogaster*, 264 had some functional annotation (2002; Hubbard, Barker et al. 2002), while 231 had more than one predicted target interaction site. Results obtained from our random model suggested that 3' UTRs with more than one predicted target site for a given miRNA are more reliable than those with a single site. We also observed multiple hits of several distinct miRNAs on a single target gene, suggesting cooperative binding and/or functional interactions between miRNAs, Figure 2.5 for examples of genes with 'brain' annotation. Other specific examples are: The eye pigmentation gene brown (bw) is hit by the miRNAs bantam (three sites) and miR-314 (two sites); the apoptosis gene hid/wrinkled (W) is hit by bantam (two sites), miR-309 and miR-286; and the eye development gene seven-up (svp) is hit by miR-33 (two sites), miR-124, miR-277 and miR-312.

We also found 150 potential targets in corresponding *A. gambiae* genes. Of these, some 40% exhibit target site conservation of more than 60% identity to the corresponding *D. melanogaster* site. Notable examples are Scr (miR-10), netrin-B (miR-184, miR-284), sticks and stones (miR-282; two hits), and VACht (miR-9).

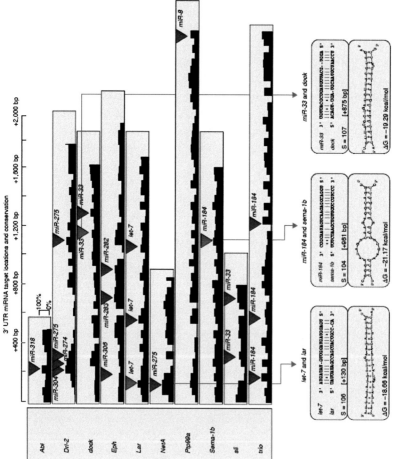

Figure 2.6 Examples of predicted sites and their conservation

Representation of 3' UTRs for potential miRNA target genes involved in axon guidance. Each individual conserved hit between a miRNA and a target gene is marked by an annotated triangle on a conservation plot (D. melanogaster versus D. pseudoobscura) for that UTR. Red triangles indicate target site locations that are illustrated in more detail (alignment and secondary structure) below. Multiple target sites on a 3' UTR for one or more miRNAs are not uncommon and reflect cooperative regulation of transcription.

2.4.4 Predicted processes and pathways targeted by Drosophila microRNAs

Our initial predictions led us to conclude that the following processes were over-represented by detected miRNA targets: transcriptional control, translational control, cell-adhesion, enzyme regulation and apoptosis regulation (Figure 2.10).

GO molecular function class

	Transcription regulator	Apoptosis regulator	Cell adhesion molecule	Binding	Enzyme	Transporter	Signal transducer	Translation regulator	Motor	Enzyme regulator	Structural molecule	Chaperone
Actual miRNAs	113	6	10	146	128	35	66	4	6	9	6	1
Random miRNAs (average)	43	1	2	98	88	22	41	2	4	9	7	1
Standard deviation	± 7.5	± 0.9	± 1.5	± 13.3	± 13.7	± 4.9	± 9.7	± 1.3	± 2.2	± 3.9	± 2.6	± 1.2
Z-score	9.3	5.9	5.0	3.6	2.9	2.7	2.6	1.4	1.0	0.01	-0.3	-0.3

Table 2.6 Functional enrichment of actual versus random microRNAs
Integers are number of detected conserved cases in each class. The standard deviation is for 50 experiments. The Z-scores for seven functional classes indicate over-representation for actual miRNA target genes. The thresholds used are S: 100; ΔG: -19 kcal/mol; ID: 70%.

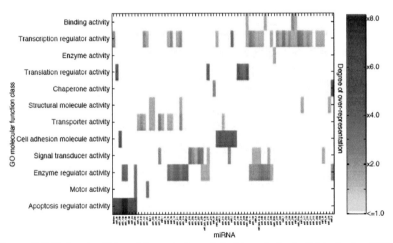

Figure 2.7 Molecular GO enrichment of predicted targets for individual microRNAs.
Functional map of miRNAs and their target genes. Left axis: selected over-represented Flybase [49] derived GO [87] classifications from the 'molecular function' hierarchy. Bottom axis: ordered list of the 73 miRNAs. Each cell in the matrix is color-coded according to the degree of over-representation (right axis) for a miRNA hitting a specific functional class. For example, a bright red box indicates that a given miRNA hits six to eight times more targets in a particular class then one would expect by chance. The matrix is built by two-dimensional hierarchical clustering after normalization for classes that are over-represented in Flybase annotations as a whole.

Separate groups of miRNAs appear to be specific to particular functional classes of target genes: for example, a group of seven miRNAs, miR-281, miR-311, miR-79, miR-92, miR-305, miR-131 and miR-31a, enriched two to four times in target genes in larval development; a group of five, bantam, miR-286, miR-309, miR-14 and miR-306, enriched three to six times in targets implicated in death or cell death; and the group bantam, miR-286 and the miR-2/miR-13 family, enriched five to six times in genes involved in regulation of apoptosis. A group of nine miRNAs is also two to three times enriched in genes involved in pattern speci-fication. Overall, target genes annotated as transcription factors are detected twice as frequently as expected by chance (21% of annotated identified target genes, compared to 9.5% of all annotated FlyBase(2002) genes). Translation factors are increased four times over expectation (miR-318, miR-304, miR-276b). This may represent a more general control mechanism for miRNA regulation of translation, in addition to the specific control of translation of individual target genes.

Investigating possible connections between genomic location and function, we analyzed 12 clusters of miRNAs in the *D. melanogaster* genome the members of which are potentially co-expressed, for example, let-7, miR-125 and miR-100 (Lagos-Quintana, Rauhut et al. 2001; Aravin, Lagos-Quintana et al. 2003; Lai, Tomancak et al. 2003; Sempere, Sokol et al. 2003). Contrary to expectation, we did not find any obvious links between genomic location and predicted target gene. One possible exception is the link between the position of three of the five copies of the miR-2 family in the intron of the gene *spitz* (involved in growth) and one of its top target genes, *reaper* (involved in cell death).

The well-characterized miRNA let-7 (in *C. elegans*) has two annotated top ten targets: *tamo* and *lar*. The gene *tamo* is thought to be required for the nuclear import of the NF-kB homolog Dorsal and recent work has connected it to the expression of a small RNA regulated by ecdysone (Minakhina, Yang et al. 2003). We do not predict *hunchback* as a target of let-7 *in D. melanogaster*, given the thresholds and parameters used, although some below-threshold hits do appear to be conserved in *D. pseudoobscura*. Instead, we predict miR-12 and miR-184 to have a stronger effect on *hunchback*.

Many genes involved in the maternal genetic system, determining germ cell fate and anterior-posterior polarity of the egg, are well known to be translationally regulated. We clearly predict the following subset of these genes to be miR-NA targets: germ cell less, bicoid, hunchback, caudal, staufen, arrest (bruno-1) and bruno-2. In addition, although the genes oskar and nanos are not top-ten ranked predictions, oskar has two conserved target sites (miR-3, miR-6), ranked below the top-ten hits for these miRNAs, and nanos has five strongly predicted target sites for miR-9c (three) and miR-263b (two), but below the 80% conservation threshold. Taken together, these data may indicate that this system is at least partially under miRNA regulation. We detail below three more biological processes predicted to be subject to miRNA regulation.

2.4.5   microRNAs targeting Hox genes and body axis specification

A multi-tiered hierarchy of transcription factors establishes the morphological segmentation and diversification of the anterior-posterior body axis of the *Drosophila* embryo (Akam 1998). The Hox genes (lab, pb, Dfd, Scr, Antp, Ubx, abd-A and Abd-B) play a key role in diversification by switching the fates of embryonic segments between alternative developmental pathways (McGinnis, Kuziora et al. 1990). The genes are organized in two separate clusters on chromosome 3R, the Antennapedia (lab, pb, Dfd, Scr, Antp) and Bithorax (Ubx, abd-A, Abd-B) complexes. Both the genes and their relative order within the complexes are conserved in vertebrates (Mann 1997).

Potential miRNA targets of Hox cluster genes and their regulators

Table 2.4
Potential miRNA targets of Hox cluster genes and their regulators

Gene name	Genes identifier	MiRNA
Abd-A	CG10325*	*miR-263a*
Abd-B	CG11648	*miR-3, miR-5, miR-306*
Antp	CG1028*	*miR-304*
Ftz	CG2047	*miR-125*
Hth	CG17117	*miR-276a, miR-265b, miR-279, miR-287†*
Pc	CG32443	*miR-100, miR-313*
Scr	CG1030	*bantam, miR-10†, miR-125, miR-315†, miR-275*
Trr	CG3848*	*miR-124*
Trx	CG8651	*miR-283, miR-307*
Ubx	CG10388	*miR-280, miR-315, miR-316*, miR-317**

*Target gene based on top 20 hits of each miRNA; all others are based on top ten hits. †Target site also conserved in *A. gambiae*.

Our predictions indicated that five of the eight Hox genes are regulated by miR-NAs. The 3' UTR of Scr is a potential target of miR-10, which is located within the Antennapedia complex between Dfd and Scr (and similarly, near the homologs of Dfd (hox4) in the Hox gene clusters of A. gambiae, Tribolium castaneum, zebrafish, pufferfish, mouse and human). Scr is also a strong target for bantam, the miRNA associated with the apoptosis gene hid (Stark, Bushati et al. 2008), and for miR-125, the putative *Drosophila* homolog of the miRNA lin-4 in *C. elegans*. Another strong target match for miR-125 is ftz, which is involved in the regulation of Hox genes and lies within the Hox cluster between Scr and Antp. All three of the Bithorax complex genes are likely to be regulated by multiple miRNAs. Interestingly, abd-A and Abd-B are both targeted by miR-iab-4-3-p, which is located within the complex between abd-A and Abd-B.

Aside from the Hox genes themselves, several other regulators of Hox gene function also appear to be miRNA targets. These include members of the trithorax activator (trx, trr) and the Polycomb (Pc) repressor groups, which control the spatial patterns of Hox gene expression by maintaining chromatin structure (Simon and Tamkun 2002), and homothorax (hth), which is required for the nuclear translocation of the Hox cofactor extradenticle (Rieckhof, Casares et al. 1997)

2.4.6   microRNAs targeting ecdysone signaling and developmental timing

Ecdysone signaling triggers and coordinates many of the developmental transitions in the life cycle of *Drosophila*. Ecdysone pulses occur during embryonic, the three larval instar, prepupal, pupal and adult stages and regulate numerous physiological processes including morphogenetic cell shape changes, differentiation and death (Riddiford, Cherbas et al. 2000; Thummel 2001; Thummel 2001) [59-62]. The regulation of these diverse processes by ecdysone is achieved through a complex genetic hierarchy. At the top of the hierarchy is the ecdysone receptor (EcR), a member of the nuclear hormone receptor family; it regulates the expression of different sets of transcription factors, including the zinc finger proteins of the Broad-Complex and many other nuclear hormone receptors, which in turn control key regulators of the different physiological processes.

We predicted many potential miRNA targets at several levels of the ecdysone cascade. These include EcR and several of the downstream transcription factors and co-factors (for example, br, E71, E74, E93, crol, fkh) (D'Avino, Crispi et al.

1995; Broadus, McCabe et al. 1999; Simon, Shih et al. 2003) Interestingly, broad (br), whose expression is exquisitely timed and differentially controlled in different tissues, has seven alternate splice forms with five different 3' UTRs.

Gene name	Gene identifier	MiRNA
Amon	CG6438	miR-2a
Aop	CG3166	miR-7§
Bon	CG5206	miR-iab-4-5p
Br‡	CG11491-RA	miR-14
	CG11491-RB&RC	miR-9, miR-14, miR-210
	CG11491-RD	let-7, miR-9
	CG11491-RE&RG	miR-9, miR-210
	CG11491-RF	miR-316
Crol	CG14938	miR-210§, miR-79
Cyp314a1	CG13478	miR-308
Eh	CG5400	miR-279
EcR	CG1765	miR-14§
Eip71CD	CG7266	miR-34
Eip74EF	CG32180	miR-306
Eip93F	CG18389	miR-14, miR-286
Fkh	CG10002	miR-281
Hr38	CG1864	miR-308
Hr46	CG33183*	miR-1 ,miR-9a, miR-9c, miR-11§, miR-124, miR-318
Hr96	CG11783†	miR-92a§
Msn	CG16973†	
Pcs	CG7761	miR-308
Rab6	CG6601	miR-317
Rpr	CG4319	miR-13a, miR-13b, miR-2a, miR-2b, miR-2c
Rep2	CG1975	miR-210
Slpr	CG22272†	miR-3
W	CG5123	bantam§
Woc	CG5965	miR-100

Table 2.5 Potential microRNA targets of ecdysone induction
* Target gene based on top 20 or 30 hits respectively for each microRNA. All others top 10 hits § Conserved in *Anopholes*

All five 3′ UTRs contain high-ranking predicted targets for miRNA regulation (Table 2.5). The fact that three miRNAs control different splice forms in varying combinations supports the analogy made to transcriptional regulation for describing the combinatorial mechanisms to achieve specificity and redundancy in targeting genes.

In addition to the core transcription factors of the ecdysone cascade, several of its effector pathways are likely to be directly targeted by miRNAs. These include genes in morphogenetic/stress signaling (aop, msn, slpr, hep), biogenesis (rab6) and the cell death pathway (hid, rpr, parcas, Rep2) (Jiang, Baehrecke et al. 1997; Jiang, Lamblin et al. 2000). They also include several miRNAs target genes that are involved in the biosynthesis of ecdysone (woc, CypP450s) and of other hormones triggering developmental transitions (amon/ETH, Eh) (Wismar, Loffler et al. 1995; Rayburn, Gooding et al. 2003). Despite their synchronous expression with ecdysone pulses in late larvae and pre-pupae, let-7 and miR-125 are not prominently targeting the core factors of the ecdysone cascade.

2.4.7   microRNAs targeting development of the nervous system

We predicted a large number of miRNA target genes that are involved in cell fate decisions in the developing nervous system. In particular, we predicted several miRNA target genes within components of the Notch pathway, which regulates the early decision between the neuronal and the ectodermal fate (Baker, Warren et al. 2000; Kadesch 2000). These include Notch ligands and factors regulating their stability (Ser, neur), as well as factors that bind to Notch (dx) or modify its sensitivity to ligands (sca, gp150, fng). They also include genes of the E(spl) complex (CG8328, CG8346) and of the Brd complex (Tom) (Lai and Posakony 1997). Genes in these two complexes are known to share motifs for translational regulation in their 3′ UTR (Bearded- and K-box), some of which have previously been predicted to be miRNA target sites (Lai and Posakony 1997). In addition, our predicted miRNA target genes include factors involved in the asymmetric cell division of neuroblasts (insc, par 6) and transcription factors regulating different aspects of neuronal differentiation (Dr, jim, Lyra, nerfin, SoxN, svp, unc4).

Some of our predictions were later experimentally validated. Our target predictions in ranked the notch signaling genes HLHm3, hairy, and HLHm4 at positions 1, 3, and 7, respectively, in the list of 143 target genes for miR-7. Experiments in *D. melanogaster* confirmed that miR-7 targets HLHm3, HLHm4, and hairy (Stark,

Brennecke et al. 2003). Similarly, our predictions ranked reaper, grim, and sickle at positions 3, 11, and 19, respectively, among the other 120 predicted target genes for miR-2b, and these predictions have now been confirmed in a collaboration with the Gaul lab (Leaman, Chen et al. 2005).

The establishment of neural connectivity is a complex morphogenetic process comprising the growth and guidance of axons and dendrites, and the formation of synapses. Many miRNAs target these processes at several different levels These targets include a remarkably large number of secreted and transmembrane factors known to mediate axon guidance decisions (netrin A and B, Slit; Drl, Dscam, Eph, PTPs, Robos, Semas; Fasl, beatlV) (Chisholm and Tessier-Lavigne 1999; Giniger 2002)[76-78]. All these factors are conserved and have similar axon guidance functions in vertebrates. Interestingly, netrin1 and robo are also predicted miRNA target genes in vertebrates (our unpublished observations), suggesting that translational control is an important conserved aspect of the regulation of axon guidance factors.

In addition to these cell surface factors, miRNAs target the cellular machinery that effects cell shape change and adhesion, including regulators and components of the cytoskeleton (for example, dock, trio, RhoGAPs, Rho1, Abl, tricornered, wasp, Sop2, nesprin, Khc-73, gamma-tubulin) and of the cell junctions (for example, crumbs, dlt, mbc, skiff).

Many of the developmental factors are re-employed in the mature nervous system to control synaptic function by effecting morphogenetic changes in synapse size, shape or strength. Additional miRNA target genes that are active in the mature nervous system include neurotransmitter receptors, ion channels and pumps (clumsy, DopR, nAChr; Hk, Shaker), as well as factors involved in neurotransmitter transport and synaptic release (SerT, vAChT, Eaat1; Cirl, Rab3, Sap47, unc13) [79].

Why would translational regulation by miRNAs feature so prominently in the development and function of the nervous system? The distances between the nucleus and dendrites/axon projections are relatively large, making nuclear regulation difficult. Furthermore, differential gene activity between and even within compartments (for example, between different portions of a growth cone or different branches of a dendritic tree) is crucial for neuronal function. Therefore, translational control near the site of action is a more efficient means of modulating gene activity than transcriptional control. For axon guidance, it has been shown that the relative abundance of adhesion molecules and chemotropic receptors on the surface of the growth cone is post-transcriptionally regulated in response to external cues presented by intermediate targets. Regulation by miRNAs would thus provide an excellent additional mechanism of post-transcriptional control.

2.5   Summary

The precise rules and energetics for pairing between a miRNA and its mRNA target sites, with probable involvement of a protein complex, are not known and cannot easily be deduced from the few experimentally proven examples. Therefore, any computational methods for the identification of potential miRNA target sites are at risk of having a substantial rate of false positives and false negatives. Based on analysis of the known examples, we have biased our method toward stronger matches at the 5' end of the miRNA, and used energy calculation plus conservation of target site sequence to provide our current best estimate of biologically functional matches. Overall, we find that conservation is a crucial filter and reduces the rate of prediction error.

Our results suggest that miRNAs target the control of gene activity at multiple levels, specifically transcription, translation and protein degradation, in other words, that miRNAs act as meta-regulators of expression control. Among biological processes, we find that the most prominent targets include signal transduction and transcription control in cell fating and developmental timing decisions, as well as morphogenetic processes such as axon guidance. These processes share the need for the precise definition of boundaries of gene activity in space and time. Our findings therefore support and expand earlier work on the role of miRNAs in developmental processes [42,80]. In addition, we predict that miRNAs also play an important role in controlling gene activity in the mature nervous system.

As miRNA and mRNA have to be present simultaneously at minimum levels in the same cellular compartment for a biologically meaningful interaction, more precise expression data as a function, for example, of developmental stage [42], will be extremely useful and will be incorporated in future versions of target prediction methods. Similarly, further work will include the analysis of potential target sites in coding regions and 5' UTRs, as well as conservation and adaptation of target sites in many species.

This genome-wide scan for potential miRNA target genes gives us a first glimpse of the complexity of the emerging network of regulatory interactions involving small RNAs (see Additional data). Both multiplicity (one miRNA targets several genes) and cooperativity (one gene is targeted by several distinct miRNAs) appear to be general features for many miRNAs, as already apparent with the discovery of the targets for *lin-4* and *let-7*. The analogy of these many-to-one and one-to-many relationships to those of transcription factors and promoter regions is tempting and elucidation of the network of regulation by miRNAs will make a major contribution to cellular systems biology. In the meantime, we would not be surprised if experiments focusing on target candidates filtered in this way have a high rate of success and help to unravel the biology of regulation by miRNA-mRNA interaction

3 Chapter Three – Human microRNA target prediction

3.1 Synopsis

This chapter describes computational prediction mammalian microRNA targets. The work was done in collaboration with the Sander lab, in 2003/4 and resulted in the publication of Human MicroRNA Targets in PLoS Biology, 2004 (John, Enright et al. 2004). As of January 2010, the paper has been cited over 700 times. Since this work was done there have been approximately 400 primary papers addressing target predictions using miRanda and least 200 reviews. We provided the miRanda web server which has been accessed more than one million times, www.microrna.org. Many of the predictions made about specific microRNA targets and the general importance in physiology and disease have been vindicated in more recent research. In the light of data generated since this paper was written and published, work is ongoing to improve the algorithm and the whole conceptual approach of miroRNA targeting. Some of this is presented in later Chapters, and some is ongoing and discussed in the concluding chapter. Supplementary Tables and material referred to in this chapter are publicly available from the PLoS Biology web site associated with the published paper.

http://www.plosbiology.org/article/info%3Adoi%2F10.1371%2 Fjournal.pbio.0020363

Published papers contained in this chapter:

Primary contributor:

Human MicroRNA targets. PLoS Biol. 2004 Nov;2(11):e363. Epub 2004 Oct 5.. PubMed PMID: 1550287
John B, Enright AJ, Aravin A, Tuschl T, Sander C, **Marks DS**.

Prediction of human microRNA targets. Methods Mol Biol. 2006;342:101-13. Review. PubMed PMID: 16957370
John B, Sander C, **Marks DS**.

Secondary contributor:
Identification of virus-encoded microRNAs. Science. 2004 Apr 30;304(5671):734-6. PubMed PMID: 15118162
Pfeffer S, Zavolan M, Grässer FA, Chien M, Russo JJ, Ju J, John B, Enright AJ, **Marks D**, Sander C, Tuschl T.

MicroRNA profiling of the murine hematopoietic system. Genome Biol. 2005;6(8):R71. Epub 2005 Aug 1. PubMed PMID: 16086853 Monticelli S, Ansel KM, Xiao C, Socci ND, Krichevsky AM, Thai TH, Rajewsky N, **Marks DS**, Sander C, Rajewsky K, Rao A, Kosik KS

mir-122, a mammalian liver-specific microRNA, is processed from hcr mRNA and may downregulate the high affinity cationic amino acid transporter CAT-1. RNA Biol. 2004 Jul;1(2):106-13. Epub 2004 Jul 1. PubMed PMID: 17179747.Chang J, Nicolas E, **Marks D**, Sander C, Lerro A, Buendia MA, Xu C, Mason WS, Moloshok T, Bort R, Zaret KS, Taylor JM

The microRNA.org resource: targets and expression. Nucleic Acids Res. 2008 Jan;36 (Database issue):D149-53. Epub 2007 Dec 23. PubMed PMID: 18158296; PubMed Central PMCID: PMC2238905.
Betel D, Wilson M, Gabow A, **Marks DS**, Sander C

3.2   Background

3.2.1   Why predict the targets of human microRNAs?

The identification of potential microRNA targets in the fly presented in Chapter 2, had three key predictions: (1) microRNAs probably target most protein coding genes in the *Drosophila* genome; (2) the microRNA: target mRNA relationship is highly combinatoric: there is a many to one and a one to many relationship between microRNAs and mRNAs, and (3) some microRNAs, and combinations of them, preferentially target specific biological processes. We wondered if there was conservation of both the scale of regulation and target identities across this evolutionary divide between insects and vertebrates. In addition we realized that insight into microRNA–target pairings might guide research in fields such as human development and cancer. Many of the predictions made in the work in this chapter have since been tested and proved consistent with some of these intial observations sin microRNA biology. Todd Golub lab showed that microR-NAs as a whole are downregulated across many different cancer type (Lu, Getz et al. 2005). This was followed by the Tyler Jacks lab showing that there ix a pronounced tansformed phenotype in mouse after down regulating key machinery in microRNA processing and a much reduced microRNA pool, (Kumar, Lu et al. 2007), see Figure 3.1.

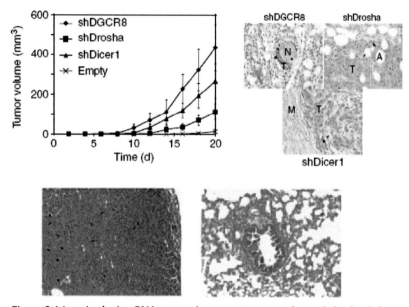

Figure 3.1 Impaired microRNA processing promotes tumorigenesis in vivo (taken from (Kumar, Lu et al. 2007)) *empty vector or shRNAs against DGCR8, Drosha or Dicer1 were injected subcutaneously into immunocompromised mice (105 cells/injection), and tumor growth was measured over time. Values are mean ±s.e.m. (n ¼ 6). (b) Hematoxylin/eosin staining of tumors, showing examples of tumor cells surrounding-nerve sheaths and infiltrating into host adipose tissue and skeletal muscle of the host. Original magnification, 40_. N, nerve sheath. A, adipose tissue. M, skeletal muscle. T, tumor cells. Arrows indicate infiltrating tumor cells. Arrowheads indicate host tissue infiltrated by tumor cell*

Bottom bit: Hematoxylin/eosin staining of a Grade III tumor from a LSL-KrasG12D Dicer1flox/+ animal. Original magnification, 20_. Black arrows indicate areas of nuclear pleomorphism. Arrowheads indicate areas with prominent nucleoli. White arrows indicate areas of vacuolation. (f) Hematoxylin/eosin staining of bronchiolar hyperplasia from a LSL-KrasG12D Dicer1flox/flox animal. Original magnification, 20.

There have now been many hundreds of papers on cancer and microRNAs (~2,500 to date), clearly implicating microRNAs as key players in the pathogenesis of tumour formation and metastasis.

3.2.2 Target sites

At the time this study was started very few microRNA targets (mRNAs) had been identified and even fewer bona fide binding sites. Hence rules for target site recognition had not been elucidated spare a small number of known pairs as described in Chapter 2. Since the work on the fly targets some new sites added some information and the following considerations about the kind of mRNA;microRNA matching were used in this work:

(1) Asymmetry: experimentally verified microRNA target sites indicate that the 5′ end of the microRNA tends to have more bases complementary to the target than its 3′ end, (Lee and Ambros 2001; Brennecke, Hipfner et al. 2003; Enright, John et al. 2003; Lewis, Shih et al. 2003; Stark, Brennecke et al. 2003)

(2) Recent experiments showed some correlation between the level of translational repression and the free energy of binding of the first eight nucleotides in the 5′ region of the microRNA (Moss, Lee et al. 1997; Doench, Petersen et al. 2003). However, confirmed microRNA:mRNA target pairs can have wobbles or mismatches in this region (Moss, Lee et al. 1997; Johnston and Hobert 2003)

(3) Loop-outs in either the mRNA or the microRNA between positions 9 and 14 of the microRNA have been observed or deduced (Lee and Ambros 2001; Brennecke, Hipfner et al. 2003; Johnston and Hobert 2003)

(4) Preferential location of G:U wobbles: wobble base pairs are less common in the 5′ end of a microRNA:mRNA duplex, and recent work shows a disproportionate penalty of G:U pairing relative to standard thermodynamic considerations (Doench and Sharp 2004).

(5) Cooperativity of binding: many microRNAs can bind to any given gene(Reinhart, Slack et al. 2000; Ambros, Lee et al. 2003; Vella, Reinert et al. 2004) and target sites may overlap (Doench and Sharp 2004).

In addition, given the overlap between the siRNA and microRNA pathways, it was reasonable to assume that rules of regulation in the siRNA pathway will offered clues to the rules applying to microRNA targeting. The following characteristics associated with siRNA functionality were identified (i) low G/C content, (ii) a bias towards low internal stability at the 3′ terminus, (iii) lack of inverted repeats, and (iv) strand base preferences (positions 3, 10, 13, and 19) (Jackson, Bartz et al. 2003; Reynolds, Leake et al. 2004). We kept these observations in mind in designing algorithms to rank target site predictions.

Since this work was done there has been a large body of work on the characterization on the microRNA:mRNA duplex. The details of the precise binding requirements for each duplex and the strength of the effect as a function of the number and position of sites are still hotly debated in the field, (Didiano and Hobert 2008).

3.2.3   Algorithms

In Chapter 2, I described the miRanda algorithm, which relies upon cross-species conservation of microRNA binding sites to identify the sites that are most likely to be functionally important. Other algorithms used to find mammalian targets include those which require a conserved seed (a consecutive run of 7/8 nucleotides), or a combination of seed and energy. This is also discussed in full in Chapter 2 and I refer the reader to the review of Sethupathy et al., (Sethupathy, Megraw et al. 2006) which has a fairly thorough assessment and description of the major programs used.

To identify binding sites in human mRNAs, we looked for conservation of binding in UTRs from the Homo sapiens, *Mus musculus* and *Rattus norvegicus* genomes. To broaden the analysis to other vertebrates, we also compared the two genomes then available for fish, the zebrafish (*Danio rerio*) and puffer fish (*Fugu ripodes*) genomes. we aimed to ask the following questions: what proportion of all genes is regulated by microRNAs? How many genes are regulated by each microRNA? Are specific cellular processes targeted by specific microRNAs or by microRNAs in general? What is the extent of cooperativity in microRNA:mRNA binding?

The Fragile X mental retardation protein, FMRP, is found in complexes with proteins such as Argonautes, which in turn bind microRNAs (Jin, Alisch et al. 2004; Plante and Provost 2006). Among prime candidates for significant microRNA control are the genes that have already been found to be posttranscriptionally regulated. The mRNA-binding protein fragile X mental retardation protein (FMRP) is involved in the regulation of local protein synthesis (Antar and Bassell 2003) and binds 4% of mRNAs expressed in the rat brain, as tested in vitro (Brown, Jin et al. 2001). The loss of function of FMRP causes fragile X syndrome, the most prevalent form of mental retardation (one in every 2,000 children). In 2004, a number of different groups had identified in vivo mRNA cargoes of FMRP. The Warren and Darnell laboratories identified ligands by co-immunoprecipitation followed by microarray analysis, complemented by extraction of polyribosomal fractions (Brown, Jin et al. 2001). They discovered that FMRP and one of its three RNA-binding domains specifically binds to G-rich quartet motifs (Brown, Jin et al. 2001; Darnell, Jensen et al. 2001; Denman 2003; Miyashiro, Beckel-Mitchener et al. 2003). Lastly, antibody-positioned RNA amplification as a primary screen followed by traditional methods identified over 80 FMRP-regulated mRNAs, with a combination of G-quartet and U-rich motifs in their mRNA sequences

(Miyashiro, Beckel-Mitchener et al. 2003; Jin, Zarnescu et al. 2004)

Independently, FMRP has been shown to be associated with RISC components and microRNAs (Caudy, Myers et al. 2002; Jin, Zarnescu et al. 2004). The *Drosophila* homolog of FMRP (FXR) and the Vasa intronic gene were identified as components of RISC (Caudy, Myers et al. 2002). Pulling together the extensive evidence from these individual studies, supported the hypothesis that the cargoes carried by FMRP are also microRNA targets.

Since the publication of the work in this chapter, it has been demonstrated that human FMRP can act as a miRNA acceptor protein for the ribonuclease Dicer and facilitate the assembly of miRNAs on specific target RNA sequences (Plante, Davidovic et al. 2006) and others have shown that FMRP regulation of mRNAs is mediated by microRNAs (Garber, Smith et al. 2006). (Yang, Xu et al. 2009)

3.3   Human microRNA targets: Methods

3.3.1   MicroRNA sequence acquisition

Mature human and mouse microRNA sequences were obtained from the RFAM microRNA registry (Griffiths-Jones 2004). To cover cases of incomplete data, any mouse microRNA sequence not (yet) described in humans was assumed to be present in human, with the same sequence, and vice versa. Similarly, all mouse microRNAs were assumed to be identical and present in the rat genome. These assumptions are reasonable as sequence identity for known orthologous pairs in human and mouse is, on average, 98% (with 110 out of 146 orthologous sequences being identical). In total, 218 mammalian microRNAs were used. For human target searches, 162 native microRNA sequences were available plus 17 mouse and 39 rat microRNA sequences; for mouse, 191 native, 14 human, and 13 rat sequences; and for rat, 45 native, 159 mouse, and 14 human microRNA sequences, Figure 3.2.

Mature microRNA sequences for zebrafish and fugu were predicted starting from known human and mouse microRNA precursor sequences (Ambros et al. 2003a). Each precursor sequence was used, in a scan against the zebrafish su-percontigs (release 18.2.1) using NCBI BLASTN (version 2.2.6; E-value cutoff, 2.0) (Altschul et al. 1990), to identify a sequence segment containing the potential zebrafish microRNA. The mammalian and fish segments were then realigned using a global alignment protocol (ALIGN in the FASTA package, version 2u65; Pearson and Lipman 1988). After testing the potential fish microRNA precursors for foldback structures (Zuker 2003), the final set of 225 predicted zebrafish mi-croRNAs was selected. The same set of sequences was used for fugu.

3'UTR sequence acquisition

The Ensembl database (Birney et al. 2004) served as the source of genomic data. The Ensembl BioPerl application user interface was used to generate 3' UTR sequences for all transcripts of all genes from each genome. Some transcripts are alternatively spliced from the same gene, so the total number of genes is smaller than the number of transcripts (Table 3). When no Ensembl annotated 3' UTR sequences were available, we predicted 3' UTRs by taking 4,000 bp of genomic sequence downstream of the end of the last exon of a transcript (Table 3). If this predicted region overlapped coding sequence on either strand, we halted 3' UTR extension at that point.

3.3.2   UTR orthology and alignment

Orthology mappings between genes from different genomes were obtained using "orthologue Tables" from the EnsMart (Kasprzyk et al. 2004) feature of the Ensembl database. Pairs of orthologous UTRs were aligned with each other using the AVID (Bray et al. 2003) alignment algorithm to facilitate analysis of conservation of position and sequence of target sites. In total, 26,205 human transcripts, representing 15,869 genes, were mapped to both mouse and rat transcripts. For zebrafish, 11,442 transcripts, representing 10,909 genes, were mapped to fugu transcripts and 11,306 transcripts mapped to human transcripts (10,063 genes).

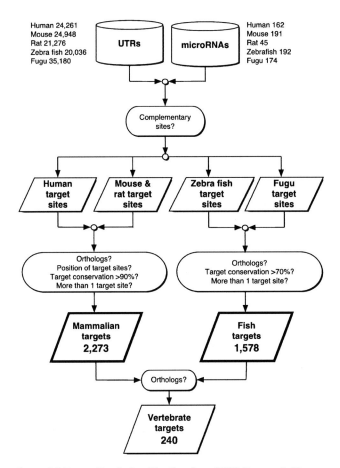

Figure 3.2 Target Prediction Pipeline for miRNA Targets in Vertebrates

The mammalian (human, mouse, and rat) and fish (zebra and fugu) 3′ UTRs were first scanned for miRNA target sites using position-specific rules of sequence complementarity. Next, aligned UTRs of orthologous genes were used to check for conservation of miRNA–target relationships ("target conservation") between mammalian genomes and, separately, between fish genomes. The main results (bottom) are the conserved mammalian and conserved fish targets, for each miRNA, as well as a smaller set of super-conserved vertebrate targets.

3.3.3   microRNA target prediction

The miRanda algorithm (version 1.2) (Enright, John et al. 2003)was used to scan all available microRNA sequences for a given genome against 3' UTR sequences of that genome derived from the Ensembl database and—tabulated separately—against all cDNA sequences and coding regions. The algorithm uses dynamic programming to search for maximal local complementarity alignments, corresponding to a double-stranded antiparallel duplex. A score of +5 was assigned for G:C and A:T pairs, +2 for G:U wobble pairs, and −3 for mismatch pairs, and the gap-open and gap-elongation parameters were set to −8.0 and −2.0, respectively. To significantly increase the speed of miRanda runs, in calculating the optimal alignment score at positions i, j in the alignment scoring matrix, the gap-elongation parameter was used only if the extension to i, j of a given stretch of gaps ending at positions i–1, j or j–1, i (but not of stretches of gaps ending at i–k, j or j, i–k for k > 1) resulted in a higher score than the addition of a nucleotide–nucleotide match at positions i, j. Removal of this restriction with the availability of more computing power would result in a moderate increase in average loop length, but the advantages of this would probably be superceded by overall refinement of target prediction rules.

Complementarity scores at the first eleven positions, counting from the micro-RNA 5' end, were multiplied by a scaling factor of 2.0, so as to approximately reflect the experimentally observed 5'–3' asymmetry; for example, G:C and A:T base pairs contributed +10 to the match score in these positions. The value of the scaling factor at each position is an adjustable parameter subject to optimization as more experimental information becomes available. Target genes (a total of 490) that contained target sites with more than one G:U wobble in the 5'end are flagged in Supplementary Table 4.4.

The thresholds used to identify candidate target sites were S > 90 and $\Delta G < -17$ kcal/mol, where S is the sum of single-residue-pair match scores over the alignment trace and ΔG is the free energy of duplex formation from a completely dissociated state, calculated using the Vienna package as in Enright et al. (Enright, John et al. 2003) (see Chapter 2).

After finding optimal local matches above these thresholds between a particular microRNA and the set of 3' UTRs in each genome, we asked whether target site position and sequence for this microRNA were conserved in the 3' UTRs of orthologous genes, i.e., between human and mouse or rat, or between fugu and zebrafish. The alignments of target sites were generated transitively (UTR microRNA UTR) via a shared (or homologous) microRNA. We required that the positions of pairs of target sites in two species fall within ±10 residues in the aligned 3' UTRs. Conserved target sites with sequence identity of 90% or more (human versus mouse or rat) and 70% or more (zebrafish versus fugu) were selected as candidate microRNA target sites and stored in a MySQL database. Using human as the reference species, we predicted 10,572 conserved target sites (conserved in either mouse or rat) in 4,463 human transcripts, of which 2,307 transcripts of 2,273 genes contained more than one target site. Similarly, using zebrafish as a reference species, we predicted 7,057 conserved target sites (conserved in fugu) in 4,820 zebrafish transcripts.

To focus on the strongest predictions, conserved target sites for each microRNA were sorted according to alignment score, with free energy as the secondary sort criterion. In cases where multiple microRNAs targeted the same site on a transcript (or within 25 nt of a site), only the highest scoring, lowest energy microRNA was reported for that site.

3.3.4   Functional analysis of predicted microRNA targets

To facilitate surveys of target function and analysis of functional enrichment, InterPro domain assignments (Mulder, Apweiler et al. 2003) and GO (molecular function hierarchy) mappings(Ashburner and Lewis 2002) for all human genes were obtained using EnsMart, www.ensembl.org. For each functional class derived from either source, we calculated its degree of under- or overrepresentation, Fclass, using the log-odds ratio of the fraction of annotated target genes with the same class (F1) and the fraction of all annotated Ensembl human genes with that class (F2):

$$F_{class} = \log_2\left(\frac{F_1}{F_2}\right), where\ F_1 = \frac{N_{tar}^{class}}{\sum_{i=1}^{i=C} N_{tar}^i}\ and\ F_2 = \frac{N_{all}^{class}}{\sum_{i=1}^{i=C} N_{all}^i}$$

Here, N represents the number of genes of a given functional class for either target genes (Ntar) or all genes (Nall), and C represents the total number of

functional classes. To eliminate bias from small counts we did not report assignments that were present in less than 1% of all annotated target genes, i.e. where
$F_1 \leq 0.01$ or $F_2 \leq 0.01$.

3.3.5   Validation

3.3.5.1 Known target sites

We sought both direct and indirect evidence to help validate or invalidate the
proposed set of mammalian targets, as follows. (1) We compared predicted targets with experimentally verified targets in mammals, *C. elegans* and *D. melanogaster*, as well as their mammalian homologs. (2) We compared predicted
target numbers from real and shuffled microRNA sequences and estimated the
rate of false-positive predictions (c.f. Chapter 2). (3) We assessed the enrichment of microRNA targets in mRNAs that are known cargoes of FMRP, an RNA-
binding protein known to be involved in translational regulation.

Indirect validation comes from the prediction that mammalian orthologs of
some of the known microRNA targets in *C. elegans* and *D. melanogaster* are
microRNA targets. Additional experimental target identification data have become available since the original publication, and these are now listed in full in
Supplementary Table 4.1.

3.3.5.2 Estimate of false positives

As a computational control of the validity of the prediction method, one can
perform a statistical test that attempts to estimate the probability that a predicted site is incorrect. Here, a "false positive" is a predicted target site of a
real microRNA on a real mRNA that has passed all relevant thresholds but is
incorrect in that it is not biologically meaningful. The statement "not biologically
meaningful" is rarely clearly defined, but can reasonably be taken to mean that
no functionally effective microRNA:mRNA interaction occurs under conditions
of co-expression at physiological concentration, where "functionally effective"
is defined in terms of detectable changes of phenotypic attributes.

An estimate of the false-positive rate can be obtained by computing (directly or
via randomization) the background distribution of scores for biologically non-
meaningful microRNA target sites and then deriving the probability that a non-

meaningful target site passes all score thresholds, i.e., for a single aggregate score, that the incorrect site has a score T > Tc, where Tc is a fixed threshold that may be different for each microRNA. We chose to estimate the background distribution using shuffled microRNAs obtained by swapping randomly selected pairs of bases of each given microRNA 1,000 times, keeping the nucleotide composition constant. The shuffled microRNA sequences were scanned against human, mouse, and rat 3' UTR sequences exactly as for the prediction procedure for real microRNA sequences. In the procedure, a microRNA:mRNA match site is predicted to be a target site if it passes three thresholds, S > Sc for match score, $|\Delta G| > |\Delta Gc|$ for free energy of duplex formation, and C > Cc for conservation, where C reflects a binary evaluation of orthology of mRNAs, similarity of position of the site on the mRNA, and a threshold percentage of conserved residues in the two mRNA target sites. Finally, the predicted target sites for a set of shuffled microRNAs are counted and then averaged over a total of ten randomized runs. The percentage of false positives for target transcripts with more than two, three, and four sites is 39%, 30%, and 24%, respectively, using a non-permissive conservation threshold of 100% for target site sequences (Figure 3.2). The false-positive rate for single sites with a score of more than 110 is approximately 35%.

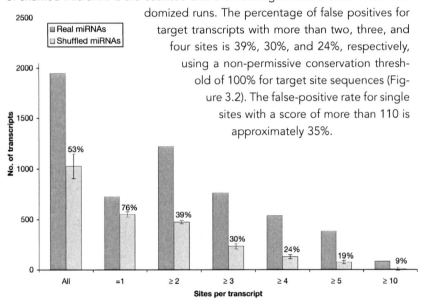

Figure 3.3 Distribution of Transcripts with Cooperativity of Target Sites and Estimated Number of False Positives. _Each bar reflects the number of human transcripts with a given number of target sites on their UTR. Estimated rate of false positives (e.g., 39% for ≥2 targets) is given by the number of target sites predicted using shuffled miRNAs processed in a way identical to real miRNAs, including the use of interspecies conservation filter._

3.3.6   mRNA s associated with RNA binding protein FMRP

We compiled a list of 464 gene identifiers of FMRP-associated mRNAs from five different publications (Brown, Jin et al. 2001; Chen, Yun et al. 2003; Denman 2003; Miyashiro, Beckel-Mitchener et al. 2003; Waggoner and Liebhaber 2003); see Supplementary Table 4.2. Among the 464 gene identifiers, 397 identifiers were mapped to the corresponding genes in our 3' UTR dataset. The remaining 67 genes were not mapped because their published identifiers were obsolete, primarily because of their Affymetrix probeset identification numbers. To identify microRNA regulation of the 397 FMRP-associated mRNAs, these genes were then compared with the set of predicted microRNA targets.

3.4   Results

3.4.1   Prediction of human microRNA targets

Using currently known mammalian miRNA sequences, we scanned 3' untranslated regions (UTRs) from the human (Homo sapiens), mouse (Mus musculus), and rat (Rattus norvegicus) genomes for potential target sites. The scanning algorithm was based on sequence complementarity between the mature miRNA and the target site, binding energy of the miRNA–target duplex, and evolutionary conservation of the target site sequence and target position in aligned UTRs of homologous genes. We identified as conserved across mammals a total of 2,273 target genes with more than one target site at 90% conservation of target site sequence (Tables S2 and S3) and 660 target genes at 100% conservation. We also scanned the zebrafish (Danio rerio) and fugu (Fugu rubripes) fish genomes for potential targets using known and predicted miRNAs (Figure 3.2 ; Tables S4 and S5) and identified 1,578 target genes with two or more conserved miRNA sites between the two fish species.

In addition to the analysis of 3' UTRs, we also scanned all protein-coding regions for high-scoring miRNA target sites. For convenience, these results are reported both as hits in cDNAs (coding plus noncoding; Table S6) and as hits in coding regions (Table S7), with cross-references in the UTR target Tables (number of hits in the coding region for each UTR in Tables S2 and S3).

The algorithm and cutoff parameters were chosen to provide a flexible mechanism for position-specific constraints and to capture what is currently known about experimentally verified miRNA target sites: (1) nonuniform distribution of the number of sequence-complementary target sites for different miRNAs; (2) 5′–3′ asymmetry (the complementary pairing of about ten nucleotides at the 5′ end is more important than that of the ten nucleotides at the 3′ end and the matches near the 3′ end can to a limited extent compensate for weaker 5′ binding); and (3) influence of G:U wobbles on binding. In choosing these parameters, we drew on experience from careful analysis of target predictions in *Drosophila* (Enright, John et al. 2003; Pfeffer, Zavolan et al. 2004)as well as proposed human targets of virus-encoded miRNAs (Pfeffer, Zavolan et al. 2004).

To facilitate evaluation of predicted targets and design of new experiments, we provide methods and results in a convenient and transparent form. We make the miRanda software freely available under an open-source license, so that researchers can adjust the algorithm, numerical parameters, and position-specific rules. We also provide web resources, including a viewer for browsing potential target sites, conserved with or without positional constraints, on aligned UTRs, with periodic updates, www.microrna.org, as well as links to these targets from the miRNA registry site RFAM (www.sanger.ac.uk). We provide both high-scoring targets, as strong candidates for validation experiments, and lower-scoring targets, which may have a role in broader background regulation of protein dose. Expression information (see Table S3) for miRNAs and mRNAs provides an additional filter for validation experiments, in addition ranking target sites by complementarity and evolutionary conservation

3.4.2 Validation of Target Predictions

3.4.2.1 Agreement with known targets.

Only a small number of target sites of target genes regulated by miRNAs have been experimentally verified, so we sought direct and indirect evidence to help validate or invalidate the proposed set of mammalian targets. (1) We compared predicted targets with experimentally verified targets in mammals, *C. elegans*, and *D. melanogaster*, as well as their mammalian homologs. (2) We compared predicted target numbers from real and shuffled miRNA sequences and estimated the rate of false-positive predictions. (3) We assessed the enrichment of miRNA targets in mRNAs that are known cargoes of FMRP, an RNA-binding protein known to be involved in translational regulation.

We previously used known miRNA sites for the *let-7* and *lin-4* miRNAs in *Drosophila* to develop the target prediction method and check for consistency (Enright et al. 2003). More recent experimental target identification provides independent control data. Recent work in *C. elegans* (Vella, Reinert et al. 2004) has narrowed the originally reported list of six target sites for *let-7* in the UTR of *lin-41* down to three elements, two target sites, and a 27-nt intervening sequence (a possible binding site for another factor). The surviving two target sites have high alignment scores, $S = 115$ and $S = 110$, while the other four sites are below threshold (Enright et al. 2003), fully consistent with the experimental results. As one of the confirmed sites has a single-residue bulge, target prediction methods that require a perfect run of base pairs near the 5′ end of the miRNA would not detect it, while our method does. *lsy-6*, a recently experimentally identified miRNA in *C. elegans*, controls left–right neuronal asymmetry via *cog-1*, an Nkx-type homeobox gene; the *cog-1* gene has a target site in its 3′ UTR, which also has a high score ($S = 125$) and passes the conservation filter.

Experiments in *D. melanogaster* have identified six new miRNA–target gene pairs: *miR-7* targets the notch signaling genes *HLHm3*, *HLHm4*, and *hairy*, and *miR-2b* targets the genes *reaper*, *grim*, and *sickle* (Stark, Brennecke et al. 2003). Consistent with these experiments, our target predictions in *D. melanogaster* (Enright, John et al. 2003)ranked *HLHm3*, *hairy*, and *HLHm4* at positions 1, 3, and 7, respectively, in the list of 143 target genes for *miR-7*. Similarly, our predictions ranked *reaper*, *grim*, and *sickle* at positions 3, 11, and 19, respectively, among the other 120 predicted target genes for *miR-2c*. We also predicted *miR-6* to target this group of pro-apoptotic genes, with sites that have lower scores than the *miR-2* family but are conserved in *D. pseudoobscura*. Unfortunately, one cannot necessarily use these validated target sites alone for the derivation of new prediction rules, as they may produce large sets of false negatives which are a result of biased subsets which were tested.

Indirect validation comes from the prediction that mammalian orthologs of some of the known microRNA targets in *C. elegans* and *D. melanogaster* are microRNA targets. An example is the proposed conservation of the microRNA–target relationship lin-4:lin-28 (we use the notation microRNA:mRNA for a microRNA–target pair), first discovered in worm (Moss and Tang 2003): we detect target sites in human lin-28 for the lin-4 microRNA homolog miR-125. We also confirm the human analog of a let-7:lin-28 relation predicted in *C. elegans* (Reinhart, Slack et al. 2000). In summary, the predicted target sites on human lin-28 are miR-125 (1 site), let-7b (2 sites; (Moss and Tang 2003)), miR-98 (2 sites), and miR-351 (1 site). Another known lin-4 and let-7 target in *C. elegans* is lin-41. The human homolog of lin-41 and another closely related gene (encoding Tripartite motif protein 2) are predicted as high-ranking targets of let-7 and miR-125 (the human homolog of lin-4) (see Supplementary Tables 2 and 3). Another known instance of microRNA target regulation in worms is the regulation of cog-1 by the lsy-6 microRNA (Johnston and Hobert 2003). Although there is no obvious homolog of lsy-6 in mammals, the vertebrate homolog of the target gene cog-1, nkx-6.1, is a conserved target for five different microRNAs in our predictions (see Supplementary Table 4.1).

The comparison of our results with known targets shows that our method can detect most (but not all) known target sites and target genes at reasonably high rank. However, given the small number of experimentally verified miRNA–target pairs, additional validation tests are desirable, such as statistical tests using randomization of miRNA sequences to estimate false positives.

3.4.2.2 Estimate of false positives.

As a computational control of the validity of the prediction method, one can perform a statistical test that attempts to estimate the probability that a predicted site is incorrect. Here, a "false positive" is a predicted target site of a real miRNA on a real mRNA that has passed all relevant thresholds but is incorrect in that it is not biologically meaningful. The statement "not biologically meaningful" is rarely clearly defined, but can reasonably be taken to mean that no functionally effective miRNA:mRNA interaction occurs under conditions of co-expression at physiological concentration, where "functionally effective" is defined in terms of detectable changes of phenotypic attributes.

Technically, an estimate of the false-positive rate can be obtained by computing (directly or via randomization) the background distribution of scores for biologically non-meaningful miRNA target sites and then deriving the probability that a non-meaningful target site passes all score thresholds, i.e., for a single aggregate score, that the incorrect site has a score $T > T_c$, where T_c is a fixed threshold that may be, in general, different for each miRNA. We chose to estimate the background distribution using shuffled miRNAs obtained by swapping randomly selected pairs of bases of each given miRNA 1,000 times, keeping the nucleotide composition constant. The shuffled miRNA sequences were scanned against human, mouse, and rat 3′ UTR sequences exactly as for the prediction procedure for real miRNA sequences. In the procedure, a miRNA:mRNA match site is predicted to be a target site if it passes three thresholds, $S > S_c$ for match score, $|\Delta G| > |\Delta G_c|$ for free energy of duplex formation, and $C > C_c$ for conservation, where C reflects a binary evaluation of orthology of mRNAs, similarity of position of the site on the mRNA, and a threshold percentage of conserved residues in the two mRNA target sites. Finally, the predicted target sites for a set of shuffled miRNAs are counted and then averaged over a total of ten randomized runs. The percentage of false positives for target transcripts with more than two, three, and four sites is 39%, 30%, and 24%, respectively, using a non-permissive conservation threshold of 100% for target site sequences (Figure 3.2). In addition, the false-positive rate for single sites with a score of more than 110 is approximately 35%.

To provide a realistic estimate of false positives using randomization, the distribution of scores from random trials ("random-false") should be similar to the distribution of incorrect (non-meaningful) hits from real trials ("real-false"). The difference between these two distributions is difficult to compute in principle, as very few validated correct predictions are known at present. For human sequences, without any conservation filter, we obtained a total of 2,538,431 predicted target sites for real miRNAs, and, for shuffled miRNAs, on average, 2,033,701 (± 82,172) target sites—a difference of 20%. This difference may be indicative of a biological signal in the raw score *(S)* and energy *(ΔG)* calculated by the miRanda algorithm or may be due to different polynucleotide compositions of shuffled miRNAs compared to real miRNAs. Even if this difference represents a real effect, by far the most predictive criterion for accurate target detection is conservation of target sites across species, and not alignment scores or energies (20% compared to a factor of three, see Figure 3.3; Table S8). As a consequence, the current set of predicted targets rests heavily on the criterion of conservation of miRNA:mRNA match between different species. We believe this to be essentially true for all currently published target prediction methods.

3.4.2.3 Indirect experimental support: FMRP-associated mRNAs.

An excellent opportunity to test our target predictions comes from experiments showing the association of mRNAs and miRNAs with proteins involved in translational control, even if these experiments do not provide information on specific miRNA:mRNA pairings. In particular, FMRP, which may regulate translation in neurons, not only associates with hundreds of mRNAs with miRNAs (Brown, Jin et al. 2001; Darnell, Jensen et al. 2001; Denman 2003; Miyashiro, Beckel-Mitchener et al. 2003). but also associates with components of the miRNA processing machinery, *Dicer*, and the mammalian homologs of AGO1 and AGO2 (Jin, Zarnescu et al. 2004). If all FMRP-bound mRNAs are regulated by miRNAs, one should see a large enrichment of predicted targets among such mRNAs, Figure 3.4.

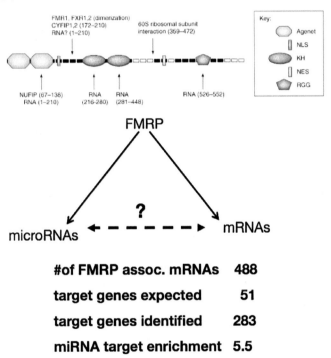

#of FMRP assoc. mRNAs   488

target genes expected   51

target genes identified   283

miRNA target enrichment   5.5

Figure 3.4 FMRP domain structure and hypothesis
FMRP has been associated with both microRNAs and sets of mRNAs; we use this as evidence that FMRP bound mRNAs might be enriched for predicted microRNA sites

We tested this hypothesis with 397 FMRP-associated mRNAs taken from a number of recent experiments.

Are FMRP-bound messages enriched in predicted targets? Using five different datasets (Table S9), we predicted that 74% of FMRP-associated messages are miRNA target genes (294 of 397 mRNAs). This corresponds to an enrichment factor of about five compared to the 59 targets one would expect from our analysis in a randomly chosen set of 397 mRNAs, where 59/397 equals 4,462/29,785 (4,462 predicted mammalian target mRNAs pass the 90% conservation filter for one or more sites per transcript out of a total of 29,785 transcripts). This suggests that in the 397 FMRP target genes, 59 should pass the filters. The enrichment factor does not vary much with the cutoff parameters used in target prediction (data not shown), but is subject to some uncertainty because of potential false-positive predictions. The enrichment of miRNA:FMRP interaction is consistent with the hypothesis that translational control involving FMRP protein is executed in a complex that involves one or more miRNAs interacting with transcripts at specific sites. Note that this analysis supports the validity of target gene prediction, not the identity of the controlling miRNA or the accuracy of specific sites.

An additional validation test involved FMRP cargoes that had been identified in more than one study, using independent experimental methods. For example, the mRNAs of 14 genes (Brown, Jin et al. 2001; Weiler, Spangler et al. 2004) were overrepresented in both the polyribosome fraction of mouse fragile X cells and in co-immunoprecipitation with mouse brain mRNPs that contain FMRP. Almost all of the 14 genes are predicted targets with more than one conserved site (11 of 12 annotated UTRs; Table S9). In some cases, expression data provide additional support: *postsynaptic density protein 95 (PSD95)–associated (SAPAP4)*, a neuron-specific protein, is regulated by many miRNAs highly expressed in rat brain primary cortical neurons (Weiler, Spangler et al. 2004).

In summary, the three validation approaches (retrospective, statistical, and indirect experimental) suggest that the current version of the miRanda algorithm, in spite of clear limitations, can predict true miRNA targets at reasonable accuracy, provided that (1) the targets are detected as conserved and (2) the gene contains more than one miRNA target site or a single high-scoring site ($S > 110$, approximately, including sites with almost perfect complementarity suggestive of mRNA cleavage).

3.5   Overview of mammalian microRNA target genes

3.5.1   More than 2,000 mammalian targets.

We predicted 2,273 genes as targets with two or more miRNA target sites in their 3' UTRs conserved in mammals at 90% target site conservation (see Tables S2 and S3). This means we predicted approximately 9% of protein-coding genes to be under miRNA regulation. In addition, we predicted another 2,128 genes with only one target site, but the false-positive rate for these is significantly higher (Figure 3.2). Of these, the top-scoring 480 genes ($S > 110$) have an estimated false-positive rate comparable to that of genes with multiple sites and thus also are good candidates for experimental verification.

Some of the genes with single sites may contain additional sites that we cannot detect for a number of reasons, including truncated UTRs. A significant subset of the total number of single-site target genes (7%) has near complementary single sites. These near complementary sites may indicate cleavage, for which additional sites may not be necessary. The targets listed in Table 3.1 were selected for variety of function, variation in number of sites, and varied extent of conservation (some are also conserved in fish).

3.5.2   One-to-many and Many-to-one

Somewhat surprisingly, the number of predicted targets per miRNA varies greatly, from zero (for seven miRNAs) to 268 (for *let-7b*), but the distribution is nonuniform (mean = 7.1, standard deviation = 4.7; Figure 3.5). This indicates a range of specificity for most miRNAs and suggests that regulation of one message by one miRNA is rare.

3.5.3   Functional analysis.

We analyzed the distribution of functional annotation for all targets of all miRNAs using Gene Ontology (GO) terms (see Materials and Methods; Table S10) and domain annotations from InterPro (Mulder et al. 2003). The target genes reflected a broad range of biological functions, Figure 3.6. The most enriched GO term was "ubiquitin-protein ligase activity," with 3.3-fold enrichment (Table S10). Since ubiquitination is a process controlling the quantity of specific pro-

Table 3.1

Target			miRNA
Gene	Identifier	Description	ID
BNOP3L	104765	Bcl2/Adenovirus E1b protein-interacting protein-3-like protein	miR-17
CASP2	106144	Caspase 2 precursor	**miR-181b**, miR-199a
	112577	Delta1	miR-17-5p, miR-363, miR-351
DICER1	100697	Endoribonuclease Dicer	miR-107, let-7e
EIF4E	151247	Eukaryotic translation initiation factor 4E	miR-17-5p, miR-206 miR-325, miR-23,
EIF4EBP3	131503	Eukaryotic translation initiation factor 4E binding protein 3	miR-136
EFNB2	125266	Ephrin B2 ligand	miR-312, miR-217, miR-340
HES1*	114315	Transcription factor HES-1	**miR-30**, miR-24
HOX-C8	037965	Hox-3A, Hox-3.1	**miR-196**, let-7b,c,d
HOX-D8	175879	HOX-D8	**miR-196**, miR-203, miR-143
HOX-B8	120068	HOX-B8	**miR-196**, miR-328, miR-101b
LAR	142949	Leukocyte-antigen-related protein	miR-133, miR-341, miR-347
MACF1	127603	Microtubule–actin cross-linking factor 1	**miR-130**
MECP	169057	Methyl-CPG-binding protein 2	**miR-199a**, let-7, miR-295
MYCC	136997	C-myc	let-7b,c, miR-187
MYCN	134323	N-myc proto-oncogene protein	**miR-17**, miR-152, miR-30a
NOG	183691	Noggin precursor	**miR-152**
NBPHOX	109132	Paired mesoderm homeobox protein 2b	**miR-7**
PCTK2	59758	Serine/Threonine-protein kinase PCTAIRE-2	**miR-18**
PLXDC2	120594	Tumor endothelial marker 7-related precursor	miR-211, miR-10a
PLXNB1	164050	Semaphorin receptor	**miR-130b**, miR-138, miR-130b, miR-245, miR-245
SARA1	79332	COPII-associated small GTPase	**miR-106**, miR-17-5p, miR-20, miR-203
SLC7A1	139514	CAT-1/High affinity cationic amino acid transporter	**miR-122a**
SMARCD2	108604	Swi/Snf-related matrix-associated regulator of chromatin d2	miR-30b, miR-234, miR-206, miR-206
SOS2	100485	Son of sevenless protein homolog 2	**miR-98**, let-7i,7e, miR-103 miR-107, miR-134
S77	004866	Suppression of tumorigenicity 7 isoform A	**miR-301**, miR-302, miR-266, miR-151
UBE3A	114062	Ubiquitin-protein ligase E3a	**miR-184**, miR-103 miR-140 miR-17_5p
USP46	109189	Ubiquitin-specific protease 46	**miR-190**
N.A.	128594	NAG14 protein	miR-33, miR-150, miR-300, miR-99b, miR-231
N.A.	142864	PAI-1 mRNA-binding protein	**miR-30**, miR-19a

Add "ENSG00000" to the beginning of the identifiers to derive Ensembl identifiers. All miRNA–target relationships shown here are conserved in mammals, i.e., homologous miRNAs target transcripts of homologous genes at similar UTR positions with similar local sequence. Genes that are predicted to be targets in both mammals and fish are in bold. Where the miRNA–target relationship is also conserved in non-mammalian vertebrates, the miRNA is in bold.
* Contains conserved CPE motif.
N.A. not available.
DOI: 10.1371/journal.pbio.0020363.t001

teins in a cell at specific times, miRNA regulation of components of the ubiquitin pathway could increase protein levels. Other overrepresented functional terms were "neurogenesis" (3.2-fold), "protein serine/threonine kinase" (2.5-fold), and "protein-tyrosine kinase activity" (2.5-fold).

The four domains most overrepresented in predicted targets relative to all genes were *Homeobox* domain, 5.3-fold; *KH* domain, 4.0-fold; and *Guanine-nucleotide dissociation stimulator CDC25* domain, 3.4-fold (Figure 3.6; Table S10).

Interestingly, *KH* domains are RNA-binding domains found in a wide range of proteins such as hnRNPk, FMR1, and NOVA-1. In addition to the *Homeobox* domain, other DNA-binding domains and domains associated with chromatin regulation were also enriched, suggesting that miRNAs in animals target the transcription machinery disproportionately, as they do in plants.

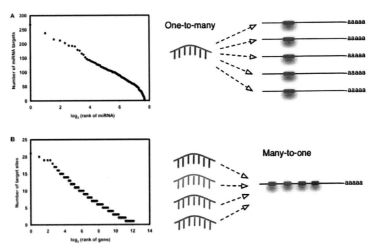

Figure 3.5 Multiplicity and Cooperativity in miRNA–Target Interactions
One miRNA can target more than one gene (multiplicity) (A), and one gene can be controlled by more than one miRNA (cooperativity) (B). The distributions are based on ordered (ranked) lists and decay approximately exponentially (approximate straight line in log-linear plot).(A) Some miRNAs appear to be very promiscuous (top left), with hundreds of predicted targets, but most miRNAs control only a few genes (bottom right). (B) Some target genes appear to be subject to highly cooperative control (top left), but most genes do not have more than four targets sites (bottom right). Although specific values are likely to change with refinement of target prediction rules, the overall character of the distribution may well be a biologically relevant feature reflecting system properties of regulation by miRNAs.

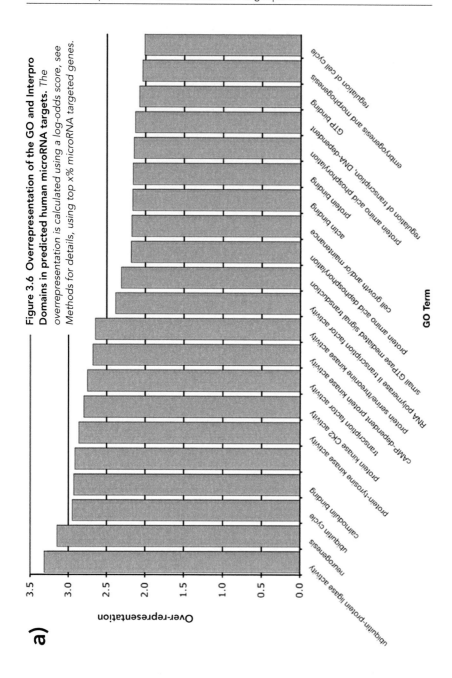

Figure 3.6 Overrepresentation of the GO and Interpro Domains in predicted human microRNA targets. *The overrepresentation is calculated using a log-odds score, see Methods for details, using top x% microRNA targeted genes.*

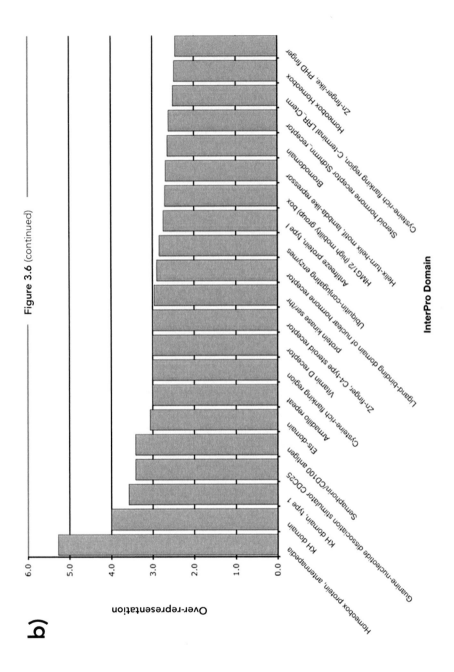

Figure 3.6 (continued)

Another overrepresented domain was *semaphorins* (3.0-fold). The *semaphorins* and *plexins* (semaphorin receptors) are involved in axon guidance, angiogenesis, cell migration, the immune system, and the adult nervous system (Pasterkamp and Verhaagen 2001). Many semaphorins and their receptors are predicted targets of brain-expressed miRNAs (e.g., *let-7c, miR-125b, miR-153, miR-103, miR-323, miR-326,* and *miR-337*). The plexins dimerize with Neuropilin (NP1) to signal the Semaphorin ligand attachment; *neuropilin* is a predicted high-ranking target of *let-7g* and *miR-130*, both brain-expressed miRNAs. A significant proportion of ephrin receptors (seven out of ten genes) and ephrin ligands (five out of seven genes) are predicted targets. The family of *ephrins* is involved in boundary formation, cell migration, axon guidance, synapse formation, and angiogenesis, and the ephrin ligand, *EphA2*, which contains a conserved cytoplasmic polyadenylation element (CPE) motif, is considered to be under translational regulation in axon growth cones (Steward and Schuman 2003). Although many members of the *ephrin* family are predicted targets of brain-expressed miRNAs, they appear to be targeted by different miRNAs, consistent with differential regulation. In *Drosophila*, both *ephrin* and *EphR*, closest to the mammalian B class of the *ephrin* family, also are predicted miRNA targets.

Do specific miRNAs target particular functional groups? We analyzed each miRNA individually for GO term and domain enrichment (Table S10). The targets of some miRNAs were strongly enriched in certain categories, e.g., *miR-105* in "small GTPase mediated signal transduction" (5-fold), *miR-208* in "transcription factor" (6-fold), and *miR-7*, which lies in the intron of the hnRNPk (an RNA-binding protein) gene, in "RNA binding proteins." Neuronal differentiation of embryonic carcinoma cells by retinoic acid in both mice and humans is coupled to induction of *let-7b, miR-30, miR-98, miR-103,* and *miR-135* (Sempere, Freemantle et al. 2004), and their targets are enriched in "neurogenesis" (3.5-fold). *miR-124a* and *miR-125*, both highly and specifically expressed in brain, preferentially target RNA-binding proteins. Thirty-one new miRNAs *(miR-322–miR-352)* cloned from rat neuronal polyribosomes have a large number of neuronal target genes and share many targets, e.g., *miR-352* and *miR-327* target 5HT-2c, and *miR-340, -328, -326, -331,* and *-333* potentially target beta-catenin, which is implicated in various stages of neural differentiation.

Two highly expressed miRNAs in the thymus, *miR-181a* and *miR-142–3p* are key components of a molecular circuitry that modulates hematopoietic lineage (Chen et al. 2004). Ectopic expression of *miR-181a* causes a 2-fold increase in the cells of the B cell lymphoid lineage. Some of our high-ranking targets for *mir-181a* may provide clues for the mechanism of this effect. Germ cell nuclear factor *GCNF (NR6A1)* (the second-highest-ranked target for *mir-181a*) is expressed in the thymus and bone marrow. *mir-181a* itself is encoded on the antisense strand of an intron of *GCNF*. We also predict that the gene *Bcl11b*, known to affect B cell growth, is a target of *mir-181a*, ranking third, as well as Lim/homeobox protein LHX9, recently found expressed in developing thymus (Woodside et al. 2004).

3.6   FMRP cargo mRNAs regulated by miRNAs

3.6.1   Overview of microRNA targeted FMRP cargo

FMRP is composed of several RNA-binding domains (two KH and one RRG) that bind messages.

Target				miRNA
Gene	Reference[a]	ID	Description	ID
APP[b,c]	Denman 2003	142192	Amyloid beta A4 protein precursor	*let-7*[d], *miR-130*, *miR-214*
BASP1[c]	Brown et al. 2001	176788	Neuronal axonal membrane protein NAP-22	*miR-207*, *miR-18*, *miR-22*
CACNA1D	Chen et al. 2003	157388	Voltage-dependent L-type calcium channel alpha-1D subunit	*miR-291–5p*
CIC[b,c]	Brown et al. 2001	079432	Capicua *(Drosophila)* homolog	*miR-202*, *miR-210*, *miR-292-as*
CLTC	Chen et al. 2003	141367	Clathrin heavy chain 1	*miR-122a*, *miR-330*
DDX5	Chen et al. 2003	108654	Probable RNA-dependent helicase P68	*miR-1d*, *miR-147*, *miR-154*, *miR-33*
DLG3[c,e]	Chen et al. 2003	082458	Presynaptic protein SAP102	*miR-15b*, *miR-196*, *miR-326*
DLG4[c]	Brown et al. 2001	132535	PSD95, presynaptic density protein	*miR-125a*[d], *miR-135*, *miR-324–3p*
FACL3[b,c]	Brown et al. 2001	123983	Long-chain Acyl-CoA synthetase 3	*let-7*[d], *miR-141*, *miR-98*
FMR1[c]	Brown et al. 2001	102081	FMRP1	*miR-194*[d], *miR-297*[d], *miR-326*
FMR2	Chen et al. 2003	155966	FMRP2	*miR-152*[d]
FXR1[b]	Denman 2003	114416	FMRP1	*let-7*[d], *miR-199*, *miR-336*
HNRPA2B[c]	Chen et al. 2003	122566	Heterogeneous nuclear ribonucleoproteins A2/B1	*miR-103*[d], *miR-143*, *miR-151*
HTR1B[b]	Denman 2003	135312	5-hydroxytrypatmine 1B receptor	*miR-292-as*, *miR-25*, *miR-202*, *miR-183*
HTR2C[c]	Denman 2003	147246	5-Hydroxytrypatmine 2C receptor	*let-7e*, *miR-352*, *miR-199-as*, *miR-9*
MAP1B	Brown et al. 2001	131711	Microtubule-associated protein 1B	*miR-325*, *miR-136*
MAP4K4	Brown et al. 2001	071054	Mitogen-activated protein kinase 4	*miR-29a*
Mint homolog[b]	Brown et al. 2001	065526	Smart/HDAC1-associated repressor protein	*miR-203*
SEMA3F	Miyashiro et al. 2003	001617	Semaphorin 3F	*miR-182*, *miR-325*

Table 3.2
Transcripts for genes (Gene and ID) are described as FMRP cargoes in several studies (DR) and predicted here as targets of specific miRNAs (MiRNA). Selected from a total of 294 such targets.
[a] Reference from which data was extracted.
[b] Homologous miRNA-mRNA pair conserved in fish.
[c] Additional miRNAs are predicted to target the gene (number in parentheses): APP (9), BASP1 (4), Capicua (2), DLG3 (7), and DLG4 (5).
[d] The miRNA has multiple target sites on the gene.
[e] The 3′ UTR of the gene contains a CPE motif (Table S11).
DOI: 10.1371/journal.pbio.0020363.t002

The specific binding motifs for FMRP on messages are incompletely known, but are thought to include G-quartet patterns and/or U-rich sequences ((Dolzhanskaya, Sung et al. 2003; Ramos, Hollingworth et al. 2003) predicted 294 mRNAs known to be FMRP cargoes as miRNA targets (see Table S9). The most reliable of these (Table 3.2) reflect high confidence in experimental identification of FMRP association or conservation of target site between mammals and fish.

3.6.2   Alzheimer's disease amyloid protein

Amyloid precursor protein (APP) is an FMRP-bound protein that is translationally regulated. The *APP* transcript contains a 29-nt motif at position 200 in the 3' UTR that is known to aid destabilization of the *APP* mRNA in certain nutrient conditions and that binds nucleolin, a protein associated with RNPs containing FMRP ((Rajagopalan and Malter 2000). In addition, there is an 81-nt sequence at position 630 in the *APP* 3' UTR that is required for the TGF beta-induced stabilization of the *APP* mRNA (Amara, Junaid et al. 1999).We predicted *APP* as a target, with a total score of $S = 708$ with a minimum of eight miRNA sites, including two *let-7* top-ranking sites that are conserved in human, mouse, and rat. One of the predicted miRNA target sites in the *APP* 3' UTR lies in the 81-nt region (Figure 3.5), and another is within 30 nt of the motif at position 200.

Other APP-interacting proteins, APP-binding family B member 1 *(mir-9, miR-340, and miR-135b)*, APP-binding family member 2 *(let-7 and miR-218)*, and APP-binding family 2 *(miR-188 and miR-206)* were also predicted targets, some of which had near exact target site matches. In summary, the *APP* gene appears to be subject to translational regulation by the combinatorial control of a number of different miRNAs.

3.6.3   **PSD95 and synaptic processes**

PSD95 and similar scaffolding molecules, link the NMDA receptor with intracellular enzymes that mediate signaling; this process is involved in the development and maintenance of synaptic function and synaptic plasticity, and interference in this process is implicated in schizophrenia and bipolar disorder

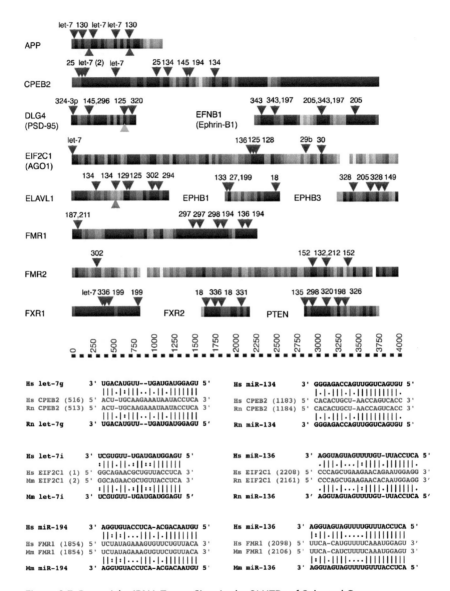

Figure 3.7 Potential miRNA Target Sites in the 3' UTRs of Selected Genes

Nucleotide sequence conservation between the 3' UTRs of human and the closest mouse or rat orthologous genes is averaged for each block of 40 base pairs (long rectangles; white indicates 0% identical nucleotides, black indicates 100% identical nucleotides, and grey indicates intermediate values). The positions of target sites for specific miRNAs (triangles above rectangles, with numbers indicating miR miRNAs, e.g. "130" is "mir-130") are, in general, distributed nonuniformly. Sequence motifs other than target sites (triangles below rectangles) are mRNA stability elements (APP), a G-quartet (DLG4), and an AU-rich element (ELAVL1), representing possible protein-binding sites. Detailed alignments between the miRNA and the predicted target sites (arbitrary selection) illustrate, in general, stronger match density at the 5' end of miRNAs than at the 3' end, as required by the algorithm and as observed in experimentally validated targets. The nonconserved nucleotides in the target sites are highlighted in red. Gene names map to the following Ensembl identifiers (142192 is ENSG00000142192, etc.): APP, 142192; CPEB2, 137449; DLG4, 132535; EFNB1, 090776; EIF2c1, 092847; ELAVL1, 066044; EPHB1, 154928; EPHB3, 182580; FMR1, 102081; FMR2, 155966; FXR1, 114416; FXR2, 129245; and PTEN, 171862.

FMRP binds PSD95 and is required for mGluR-dependent translation of PSD95(Todd, Mack et al. 2003). PSD95 is a high-ranking target of miR-125, miR-135, miR-320, and miR-327, all of which are either exclusively expressed in brain or enriched in brain tissue (Krichevsky, King et al. 2003; Lagos-Quintana, Rauhut et al. 2003; Sempere, Sokol et al. 2003). In particular, large transcript numbers of miR-125b are found copurified with polyribosomes in rat neurons in (Kim, Krichevsky et al. 2004). PSD95 has one reported G-quartet in its 3' UTR at position 648(Todd, Mack et al. 2003) (Ceman, Brown et al. 1999), further suggesting it as an in vivo FMRP target. We predicted an additional G-quartet site at position 205–235 in the 3' UTR of PSD95. One of the miRNA (miR-125) target sites overlaps with the G-quartets, raising the possibility that miRNAs directly compete with FMRP to bind the message in this location. Likewise, NAP-22, which has three miRNA target sites (see Table S9), has a miR-207 target site that overlaps with a G-quartet (Darnell et al. 2001).

Other PSD95 family members are also involved in synaptic processes, in particular, in the integration of NMDA signaling in the synaptic membrane. All PSD95 family members in mammals (also known as discs large 1–5), SAP90, and CamKII are predicted miRNA targets (see Table S9), as well as mGluR, the protein product of which is an agonist that induces the rapid translation of PSD95 (Todd et al. 2003) and three NMDA receptor subunits (see Table S9). These results suggest that miRNAs may be involved in NMDA and glutamate receptor signaling to coordinate and integrate information, with specificity achieved through the combinatorial action of different miRNAs.

3.7   Components of RNPs Regulated by miRNAs

3.7.1   FMRP-associated proteins

FMRP binds its own mRNA, implying negative feedback if the binding inhibits FMRP production (Ceman, Brown et al. 1999). The fact that miRNAs target transcripts for FMRP and FMRP-binding proteins suggests another negative feedback loop in which high levels of these proteins inhibit their own production, Figure 3.8, (depending, of course, on the concentration of miRNAs and mRNAs). The genes for six FMRP-associated (not associated at the same time) proteins, hnRNP A1, Pur-alpha, Pur-beta, Staufen, AGO-2, and PABP, are predicted miRNA targets. This indicates that FMRP-containing RNPs are under miRNA regulation. *FXR2*, a gene similar to *FMR1* is also a miRNA target in human, mouse, rat, and fish. Details of the implied feedback regulation and differential control of RNP action remain to be determined.

3.7.2   RISC

Our data suggest that the RNAi–miRNA machinery itself is under miRNA regulation, Figure 3.8.

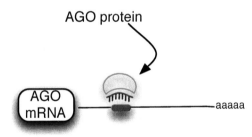

Figure 3.8 Schematic of AGO self-inhibition
When microRNAs bind to AGO they can in turn inhibit the mRNA of AGO itself

For example Dicer appears to be controlled by let-7 and miR-15b; Ago-1 by let-7 and miR-29b/c; Ago-2 by miR-138; Ago-3 by miR-138, miR-25, and miR-103; and Ago-4 by miR-27a/b. Dicer and two of the Argonautes also are predicted to be targets in both zebrafish and fugu. The let-7 sites on the 3 UTR of Dicer and Ago-1 (Figure 3.7) will accommodate most of the let-7 variants with similar scores. The variants of let-7 are expressed in a wide range of tissues and developmental stages, suggesting broad regulation of Dicer and Ago-1 by miRNAs. In contrast, the only miRNA that targets Ago-2 is miR-138, which has so far been cloned only once in the cerebellum (Hutvagner and Zamore 2002; Lagos-Quintana, Rauhut et al. 2002). The target site for mir-138 has only one mismatch at position 8; this may induce a siRNA-like cleavage of the message (Hutvagner and Zamore 2002; Doench and Sharp 2004). Ago-3 is also a top target for mir-138, with only two mismatches in its site. We suggest that some miRNAs targeting this machinery (e.g., *let-7*, *mir-27*, *mir-29*, and *mir-103*) are expressed fairly widely, while others (e.g., *mir-138* and *mir-25*) have lower and more restricted expression

3.7.3   Other RNPs

3.7.3.1 Elav proteins

The highly conserved RNA-binding proteins, ELAV-like proteins (HuR, HuB, HuC, and HuD), contain three RNA-recognition motifs, which bind AU-rich elements in 3′ UTRs of a subset of target mRNAs These AU-rich elements increase the proteins' cytoplasmic stability and increase translatability (Perrone-Bizzozero and Bolognani 2002). Experiments have identified 18 mRNAs bound to HuB in retinoic-acid-induced cells; of the 14 we were able to map unambiguously, 12 are predicted miRNA target genes: Elavl1 (known to regulate its own mRNA), Gap-43, c-fos, PN-1, Krox-24, CD51, CF2R, CTCF, NF-M, GLUT-1, c-myc, and N-cadherin (Tenenbaum et al. 2000). Three of the ELAV-like genes themselves are also targets of a large number of miRNAs (see Tables S2 and S3; Figure 3.7). This is yet another example of miRNAs predicted to target the bound messages of RNA-binding proteins and of the regulation of RNA-binding genes by miRNAs.

3.7.3.2 Cytoplasmic Polyadenylation Binding Proteins Regulated by miRNAs

We predicted all four human cytoplasmic polyadenylation binding proteins (CPEBs) known in mammals as miRNA targets ranked within the top 170 target genes with 6–20 sites in their UTRs (Figure 3.7; Table S11). Indeed, CPEB2 is the highest-ranking gene of all transcripts. The orthologs to CPEB1 in fish and fly (known as *orb* in *D. melanogaster*) are also predicted as targets. CPEB is an RNA-binding protein first shown to activate translationally dormant mRNAs by regulating cytoplasmic polyadenylation in Xenopus oocytes (Hake, Mendez et al. 1998). It also regulates dendritic synaptic plasticity and dendritic mRNA transport (Mendez, Hake et al. 2000; Huang, Carson et al. 2003) and facilitates transport of mRNAs in dendrites together with kinesin and dynein in RNPs (Huang, Carson et al. 2003). CPEB binds to its target message through the CPE motif (UUUUAU), which must be within a certain distance of the hexanucleotide AAUAAA. CPEB keeps messages in their dormant state until phosphorylated, after which it activates polyadenylation (Mendez, Murthy et al. 2000), thereby activating translation or degradation (Mendez, Barnard et al. 2002). In addition, CPEB co-fractionates with the postsynaptic density fraction in mouse synaptosomes, consistent with translation of stored mRNAs in dendrites being part of the mechanism of synaptic plasticity. We have three more lines of evidence suggesting the notion that translational regulation by CPEB is linked to miRNA regulation. First, our target list and the list of genes regulated by CPEB significantly overlap. There are nine genes known to be CPEB-regulated, seven of which are predicted targets: alpha-CAMIIK, Map 2, Inositol 1, 4–5-Triphosphate Receptor type 1, Ephrin A receptor class A type 2, SCP-1, and CPEB3 (Mendez and Richter 2001). Second, CPEB is known to self-regulate in *D. melanogaster* (Tan et al. 2001). The CPEB1 homolog in fly, orb, and CPEBs in vertebrates are predicted miRNA targets. Third, the gene most correlated in expression to the CPEB homolog in *D. melanogaster* is a Piwi protein (Sting), a member of the Argonaute family (Pal-Bhadra, Bhadra et al. 2002) that is involved in translational regulation and in the RISC.

Among the predicted miRNA targets, 115 genes also contained CPE motifs, which were conserved in at least two mammals in the same positions in the UTRs and are therefore candidates for CPEB regulation (Table S11). Our predictions include HuB, HuR, Eif-4 gamma, DAZ associated protein 2, VAMP-2 (known to be posttranscriptionally regulated), Presynaptic protein SAP102, and brain-derived neurotrophic factor precursor. Taken together these data suggest that the CPEB genes, the known CPEB-regulated genes, and the predicted CPEB-regulated genes are strong miRNA target candidates and provide rich ground for experimentation.

3.8   Targets of Cancer-Related microRNAs

Deregulated expression of certain miRNAs has been linked to human proliferative diseases such as B cell chronic lymphocytic leukemia (Calin, Dumitru et al. 2002; Michael, SM et al. 2003) and colorectal neoplasia (Michael, SM et al. 2003) (Michael et al. 2003). Recent analysis of the genomic location of known miRNA genes suggested that 50% of miRNA genes are in cancer-associated genomic regions or in fragile sites (Calin, Sevignani et al. 2004). The miRNAs miR-15 and miR-16 are located within a 30-kb region at Chromosome 13q14, a region deleted in 50% of B cell chronic lymphocytic leukemias, 50% of mantle cell lymphomas, 16%–40% of multiple myelomas, and 60% of prostate cancers (Calin, Dumitru et al. 2002). Furthermore, miR-15 and miR-16 are down-regulated, or their loci lost, in 68% of B cell chronic lymphocytic leukemias (Calin, Dumitru et al. 2002). Similarly, miR-143 and miR-145 are down-regulated at the adenomatous and cancer stages of colorectal neoplasia (Michael, SM et al. 2003), and miR-155 is up-regulated in children with Burkitt lymphoma (Metzler, Wilda et al. 2004)

Our method predicted cancer-specific (by annotation) gene targets of miR-15a, miR-15b, miR-16, miR-143, miR-145, and miR-155. The target genes and their miRNA regulators are as follows: (1) CNOT7, a gene expressed in colorectal cell lines and primary tumors (Flanagan, Healey et al. 2003) (miR-15a); (2) LASS2, a tumor metastasis suppressor (miR-15b) (Pan, Qin et al. 2001); (3) ING4, a homolog of the tumor suppressor p33 ING1b, which stimulates cell cycle arrest, repair, and apoptosis (miR-143) (Nagashima, Shiseki et al. 2001; Yart, Mayeux et al. 2003); (4) Gab1, encoding multivalent Grb2-associated docking protein, which is involved in cell proliferation and survival (Yart, Mayeux et al. 2003) (miR-155; and (5) COL3A1, a gene up-regulated in advanced carcinoma (miR-145) (Tapper, Kettunen et al. 2001; Yart, Mayeux et al. 2003)

miR-16 has a tantalizing number of high-ranking targets that are cancer associated and specifically involved in the Sumo pathway There is increasing evidence that Sumo controls pathways important for the surveillance of genome integrity (Muller, Ledl et al. 2004). The first- and fifth-highest-ranked targets of miR-16 are Sumo-1 activating and conjugating enzymes, respectively. The top two single-site targets for miR-16 are an Activin type II receptor gene (TGF beta signaling) and Hox-A5, both known to be dysregulated at the level of protein expression in colon cancers. Both of these sites show near perfect complementary matching between miR-16 and the target genes (indicating possible cleavage). Both of these target genes are also targets for another cancer related miRNA, miR-15.

3.8.1   The E2F family are predicted targets of miR-17-5p

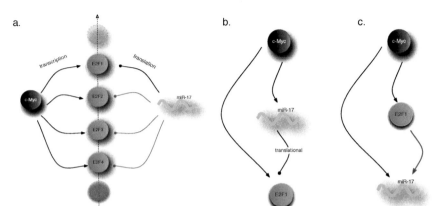

Figure 3.9 Schematic of miR-17, c-myc and E2F network
a. miR-17-5p and c-myc both regulate or predicted to regulate the E2F protein family.
b. Predicted incoherent feed-forward loop with c-myc and miR-17-5p both regulating E2F1. c. Predicted coherent feed-forward loop with c-myc and E2F1 both regulating the miR-17 cluster of microRNAs

3.9   How far are microRNA:mRNA relationships conserved in evolution?

3.9.1   Targets Conserved between Mammals and Fish

Roughly 55 miRNAs have identical mature sequences in fugu and mammals, and 80 have very similar sequences in the two species; additional fish miRNA sequences can be predicted with confidence based on sequence similarity. It is therefore reasonable to expect that the targets of these probably functionally homologous miRNAs are orthologous genes in the different species. To follow up on this hypothesis, we assessed conservation of mammalian miRNA–target pairs between the 2,273 mammalian and 1,578 fish miRNA targets (with more than one target site per UTR). The analysis yielded 240 target genes conserved between mammals and fish. The number 240 is probably an underestimate because of several factors, including: (1) unfinished annotation of genomes, particularly rat and fugu; (2) ambiguity in assigning orthologs; and (3) lack of UTR information.

The full set of conserved target genes between fish and mammals indicates a wide functional range of conserved targets (Table S12). Many Hox genes are conserved as targets, including the miR-196 targets, Hox-A4:miR-34a, Hox-C9:let-7b (near prefect complementary match), and Hox-B5:miR-27b. Examples from the notch signaling pathway include miR-30:hairy enhancer of split 1 (Hes1) and miR-152:noggin.

3.9.2   Targets Conserved between Vertebrates and Flies

Twenty-eight of the 78 identified miRNAs in flies have apparent mammalian homologs. Based on this remarkable conservation across hundreds of millions of years, it is reasonable to expect that there is some conservation of target sites, target genes, and target pathways between flies and humans. Most strikingly we can identify hox genes and axon guidance genes as common targets between vertebrates and flies, e.g., capicua and sex combs reduced (one of the vertebrate homologs of Hox-A5). The hox gene cluster in *Drosophila* contains high-ranking predicted targets (Enright et al. 2003) of miR-10 and miR-iab-4, and the hox gene cluster in mammals contains high-ranking targets of miR-196. These miRNAs are themselves located in the hox gene region. We predicted miR-iab-4–3p to target abd-B in *Drosophila*, a gene related to the ancestral hox-7 cluster, the ancestral parent of many of the predicted targets of miR-196. Axon guidance receptors and ligands conserved as targets include Lar, ephrins, and slits. Human slit1 is a top target of miR-218, which itself is transcribed from the intron of slit2, suggesting down-regulation of slit1 by transcription of slit2. We expect that there are many more conserved targets but we are hindered by the difficulty of mapping orthologous genes between human and fish. Future work will elucidate to what extent there are common pathways regulated by common miRNAs between vertebrates and invertebrates.

3.10   Investigation of "unusual" kinds of microRNA targeting

3.10.1   Target Sites in Protein-Coding Sequences

Experiments suggest that miRNA target sites in metazoans are preferentially in UTRs, not in coding regions. If this is true, a correct target site prediction method should predict a larger number of targets in UTRs than in coding regions. Alternatively, target sites in coding regions may so far have escaped experimental verification, especially in plants, in which targets of miRNAs in coding regions are the rule, not the exception.

To investigate this issue we computed the average density of target sites for high-scoring targets (S > 130) and before application of conservation filters. The statistical assessment of the influence of conservation filters in coding regions would have raised complicated issues, as nearly two-thirds of nucleotides in coding regions are conserved between mammalian genomes to preserve amino acid sequences. Interestingly, we found, on average, 11 pre-conservation target sites per 1 million nucleotides in coding regions, versus 15 such target sites per 1 million nucleotides in UTRs. This is consistent with a stronger "raw" prediction signal in UTRs and may indicate a lower number of biologically relevant target sites in coding regions in mammals, consistent with early experimental findings.

As a guide to experimentation, we report all sites in coding regions with an alignment score above 110 for miRNAs of length up to 20 nt and an alignment score above 130 for miRNAs longer than 20 nt (scores depend on the length). These cutoff scores approximately correspond to a 75% complementary match between miRNA and target, leaving open the question of how many match pairs are needed to lead to translational inhibition in coding regions, by any mechanism. We identified 942 genes that contained such sites in their coding regions. Strikingly, there was only one site with a perfect match, and this was for the imprinted miR-127, known to be antisense to the reciprocally imprinted retrotransposon-like gene on the opposite strand (Seitz et al. 2003). Of the 942 genes, 25% have been otherwise identified as targets based on conserved target sites in their UTRs. However, only five genes have targets sites in their UTRs complementary to the same miRNA that targets the coding region (see Table S3, columns H and I). For example, miR-211 has a near perfect complementary site in the coding region of a gene of unknown function (Ensembl ID ENSG00000134030, containing an Eif-4 gamma domain) and also has two conserved "normal" sites in the UTR. Similarly, miR-198 has a site in the coding region, as well as conserved sites in the UTR region, of a sodium and chloride GABA transporter (Ensembl ID ENSG00000157103). However, we see no trend for miRNAs that have conserved sites in UTRs to have additional sites in the coding region; rather, stronger target sites for a given miRNA tend to be confined either to the UTR or the coding region and are rarely in both.

3.10.2   Target Sites with near perfect matches in cDNAs

We scanned all cDNAs for high-scoring matches without using conservation to check for high-scoring targets, which we may have missed through strict conservation rules (see Table S6). Over 40 genes contain sites that have near perfect complementarity to a miRNA (S >120), and these target genes may be cleaved rather than translationally repressed as in the case of miR-196 and Hox-B8. For example miR-298, an embryonic-stem-cell-specific miRNA (Houbaviy, Murray et al. 2003) has a near match with MCL-1, and miR-328 (neuronally expressed) has a near match with LIMK-1, which is known to be involved in synapse formation and function. miR-129, expressed in mouse cerebellum, has a near perfect complementary match with Musashi-1, which is an RNA-binding gene essential for neural development, regulated in the cerebellum, and up-regulated in medulloblastoma (Yokota, Mainprize et al. 2004).

3.10.3   Comparison of miRNA Target Prediction Methods

Recently, several computational methods for the prediction of miRNA targets have been developed (Enright, John et al. 2003; Lewis, Shih et al. 2003; Stark, Brennecke et al. 2003; Kiriakidou, Nelson et al. 2004; Rehmsmeier, Steffen et al. 2004). Two of these have been applied to mammalian miRNAs, as described in Lewis et al. (2003) and Kiriakidou et al. (2004). We now compare and contrast these two methods with each other and with the current version of our method, as further developed from miRanda 1.0 and as applied to mammalian and vertebrate genomes. We compare algorithms and target lists, as an aid to the design of experiments.

The three prediction methods share the goal of identifying mRNAs targeted by miRNAs. All three use sequence complementarity, free energy calculations of duplex formation, and evolutionary arguments in developing a scoring scheme for evaluation of potential targets. Results are reported as lists of target sites and lists of target genes containing such sites. The three methods differ, however, in important technical details, such as the datasets of miRNA and UTR sequences and the algorithm and scoring scheme, as well as the report format. We now summarize these technical differences and compare the lists of resulting target genes for a common subset of miRNAs. The interpretation of such comparisons is hampered by the fact that selection criteria and the use of numerical cutoffs differ conceptually, and genomic coverage is nonuniform.

In the first method, Lewis et al. used 79 miRNAs in human, mouse, and rat, seeking targets in a UTR dataset extracted from the June 2003 version of the Ensembl database. The UTR dataset had 14,300 ortholog triplets conserved between human, mouse, and rat and 17,000 ortholog pairs between human and mouse. All annotated UTRs were extended by 2 kb of 3' flanking sequence. The algorithm required exact complementarity of a 7-nt miRNA "seed" sequence, defined as positions 2–8 from the 5' end of the miRNA, to a potential target site on the mRNA, followed by optimization of mRNA–miRNA duplex free energies between an extended window of 35 additional bases of the mRNA and the rest of the miRNA. Target genes were ranked using a composite scoring function, which took into account all sites for a particular miRNA on a given mRNA. Conserved miRNA:mRNA pairs were required to involve orthologs of miRNA and mRNA in human, mouse, and rat, but there was no requirement for conservation of target site sequence (beyond the seed match) or position on the mRNA. Using shuffled miRNA sequences, with the constraint that shuffled controls match real miRNAs in relevant sequence properties, the false-positive rate of predictions was estimated to be 50% for target genes conserved between mouse and human, 31% for target genes conserved in human, mouse, and rat, and 22% for target genes identified in fugu as well as mammals. As a final result, Lewis et al. reported 400 conserved target genes for the 79 miRNAs. Among these targets, 107 genes were reported as conserved in the fish fugu.

In the second method, Kiriakidou et al. used 94 miRNAs in human and mouse, seeking targets in a dataset of 13,000 UTRs conserved in mouse and human (from Ensembl, date not given). The algorithm used a 38-nt sliding mRNA window and calculation of miRNA–mRNA duplex free energies, keeping duplexes with energies below −20 kcal/mol. The duplexes were further filtered using a set of requirements regarding matches and loop lengths in certain positions, as derived and extrapolated from experimental tests involving a predicted target site for let-7b miRNA on the UTR of the human homolog of worm lin-28. The target site sequence was engineered into a Luciferase reporter, followed by sequence variation of the target site and test of an initial set of 15 predictions in the same reporter assay. Using shuffled miRNA sequences, and applying the same rules and parameters, the false-positive rate of predictions was estimated to be 50% for targets conserved between human and mouse. As a final result, Kiriakidou et al. reported 5,031 human targets, with 222 reported as conserved in the mouse. In the third method (this work), we used 218 mammalian miRNAs and 29,785 transcripts derived from Ensembl (Table 3) and, as a final result, report 4,467 target genes. What are the main differences between these three prediction methods? Comparison of the total number of predicted target genes is not very informative, as different datasets and cutoffs were used. We attempted to remove one of the technical differences, by explicitly comparing reported targets for the same set of 79 miRNAs used by Lewis et al. (although significant differences remained in the sets of UTR sequences used): the overlap of target genes between Kiriakidou et al. (out of 189) and Lewis et al. (out of 400) was 10.6%; the overlap between Lewis et al. (out of 400) and this work (out of 2,673) was 46%; and the overlap between Kiriakidou et al. (out of 189) and this work (out of 2,673) was 49%. In each case the totals ("out of") are the number of target genes for the common set of 79 miRNAs and the percentage is relative to the smaller set of two compared. The obvious reason for the larger overlap with our results, 46% and 49% respectively, is the larger number of targets in our predictions, which in turn is primarily the result of choice of cutoff.

Organism	Ensembl Build	Total Genes	Total Transcripts	Ensembl 3′ UTRs	Predicted 3′ UTRs
Human	19_34b	23,531	29,785	20,579	9,206
Mouse	19_32	25,329	31,387	19,496	11,891
Rat	19_3b	23,751	28,251	4,339	23,912
Zebrafish	17_2	20,036	24,469	3,495	20,974
Fugu	17_2	35,180	37,539	653	36,886

DOI: 10.1371/journal.pbio.0020363.t003

Table 3 Number of Genes and 3′ UTR Sequences Used for Target Prediction

Direct comparison of the three prediction methods is complicated by the fact that the noticeable differences between the target lists of the three methods are due to the aggregate effects of datasets, algorithm, including selection rules, use of conservation, and cutoffs. The following characteristics of the three methods underlie these differences and should be taken into consideration when choosing targets for experimentation. (1) As to UTR datasets, Lewis et al., with the earliest published report, used a smaller set of UTRs, with some likelihood of false positives as a result of UTR extension. The UTR sets used in this work, the third in terms of publication date, are the most comprehensive and plausibly the most reliable (as of February 2004). (2) As to miRNA datasets, there was an increase from 79 for Lewis et al. to 94 for Kiriakidou et al. to 218 miRNAs used in this work. (3) As to the cooperativity of binding, the scoring system of Lewis et al. evaluated cooperativity of multiple target sites by the same miRNA on a target gene, but disregarded multiple target sites from different miRNAs on one gene;

that of Kiriakidou et al. focused on single sites; and that of this work gave high scores to multiple hits on a target gene, no matter whether these hits involved the same miRNA or different miRNAs. These tendencies are not exclusive where scores involve functions of several real numbers, with cutoffs applied to the aggregate score; e.g., our method also allows strong single target sites. (4) As to assessment of false positives using statistical methods based on shuffling, the comparison of percentages is inconclusive, as the statistics of the background distribution of true negatives is not well known. It appears certain, however, from both Lewis et al. and this work, that statistical confidence increases with the extent of conservation among increasingly distant species. (5) As to validation experiments, each of the methods used a different type and set, with mixed overall conclusions. On the reassuring side, there was direct validation of some of the predicted target sites of Lewis et al. and of Kiriakidou et al. using reporter constructs in cell lines. We found some agreement between the sites validated in this way and our predicted targets (details in Table S13), but in some cases we predicted different details of target sites for a given experimentally tested miRNA:mRNA pair. Also, Kiriakidou et al. used a series of such experiments to extrapolate from a set of specific sequence variants to general rules for identification of target sites. However, serious doubts about the validity of any set of rules persist as there is very little in vivo validation in which native levels of specific miRNAs are shown to interact with identified native mRNA targets with observable phenotypic consequences under normal physiological conditions. (6) As to differences in algorithm, one can state opinions about the strengths or weaknesses of each particular algorithm, but the relationship between each prediction method and the actual in vivo process by which miRNAs have functional interactions with their target mRNAs remains unclear or, at best, unproven. In summary, in our view, each of the three methods, including the one in this work, falls substantially short of capturing the full detail of physical, temporal, and spatial requirements of biologically significant miRNA–mRNA interaction. As such, the target lists remain largely unproven, but useful hypotheses.

The predicted targets are useful in practice for the design of experiments as they increase the efficiency of validation experiments by focusing on target lists significantly enhanced in likely targets, compared to random. It is plausible that targets near the top of lists are the most likely to lead to successful experiments. Task-specific filtering of target lists for particular planned experiments is recommended, especially with respect to cooperativity of binding (more than one site for one or more miRNAs on one gene transcript) and coincidence of expression, as new data on expression patterns of miRNAs and mRNAs in different tissues become available. For example, a recommended conservative approach to the design of experiments would use all available expression information and restrict the predicted target genes to those with two or more target sites at normal threshold (S > 90) or one target site with a higher threshold (S > 110), counting only sites with up to one G:U pair in residues 2–8 counting from the 5′ end of the miRNA.

3.10.4 miRanda web server

We provided downloadable miRanda software, an interactive miRanda server for users to query their own sequences with user defined parameters and searchable, browsable target lists at www.microrna.org.

Human miRNA Targets - Search & View
April 2005 Version

- View miRNA target sites for any target gene
- View target genes and sites for any miRNA

Access to previous sets of predicted target sites is available via spreadsheets.

Although hundreds of miRNAs have been discovered in mammalian genomes, the function of less than a handful of miRNAs is known. This viewer allows access to the predicted protein coding gene targets that are conserved between human and mouse/rat. Searching by miRNA will display a ranked list of gene targets and allow you to analyze each conserved target gene/sites in detail. Similarly, searching by gene name/identifier will display the sequence of the three prime UTR and the positions of the miRNA sites along the sequence, allowing you to analyze the target sites in detail.

Download the stand-alone miRanda software for miRNA target prediction (Linux, OSX, etc)

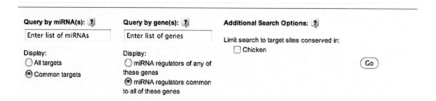

Figure 3.10 Screen shot of May 2005 version

This was updated with new target runs as new sequences became available, with improved scoring rules based on new biological information, (Betel, Wilson et al. 2008)

3.11   Discussion

3.11.1   How Widespread Is the Regulation of Translation by miRNA?

With plausible parameters, we have predicted that a *minimum* 9% (2,273 out of 23,531) of all mammalian genes have more than one miRNA target site in their 3' UTRs, with 1,314 being stronger candidates with more than two target sites. This could well be an underestimate of the total number of genes subject to miRNA regulation, as we have used a conservative conservation filter. On the other hand, not all predicted miRNA–mRNA pairs would have a biological consequence unless both miRNA and mRNA are expressed at the same time in the same cell and at sufficient concentration. The human genome has about 250 miRNA genes, compared to about 35,000 protein genes. Thus, the determination that about 1% of genes (miRNAs) control the expression of more than 10% of genes is a reasonable first order estimate. It is currently not known if any miRNAs control the expression of miRNA genes, i.e., the progression from miRNA transcript to mature miRNA.

3.11.2   How conserved in evolution aree microRNA targets?

As many miRNA sequences are detectably conserved across large evolutionary distances, they must be subject to strong functional constraints. These constraints are unlikely to come from single-site interactions with the target, as experimentally validated animal miRNAs rarely have perfectly matched target sites. Plausibly, the evolution of miRNAs is constrained by functional interactions with multiple targets. As a consequence, any compensatory mutation in the miRNA in response to mutations in a target site would be disruptive to the miRNA's interaction with other target sites. Co-evolution of the miRNA sequence and all of its target sequences is therefore a rare event. With these assumptions, the constraints on the local mRNA sequence of individual target sites are weaker than those on the miRNA sequence. We were therefore surprised to observe a substantial number of cases (28.6% of the 2,273 targets) with 100% conservation of target site sequence and with the target sites being within ten nucleotides of each other on the globally aligned UTRs of orthologous genes between mammals.

Lacking more detailed knowledge of miRNA evolution, we draw two operational conclusions. (1) Conservation of target site sequence and position is a practical information filter for predicted target sites, reducing the rate of false positives. (2) It is very likely that new miRNAs have continuously appeared in evolution (Lai 2003) at some non-negligible rate and that the set of targets for any given miRNA has lost or gained members, even between species as close as human and mouse. It is therefore important to develop prediction tools that do not rely on conservation filters or at least allow us to make them weaker. Work on this is in progress.

3.11.3   Principles of microRNA regulation - Multiplicity and cooperativity ?

Regulation by miRNAs is obviously not as simple as one miRNA–one target gene, as perhaps the early examples (lin-4 and let-7) seemed to indicate. The distribution of predicted targets reflects more complicated combinatorics, both in terms of target multiplicity (more than one target per miRNA) and signal integration (more than one miRNA per target gene).

The distribution of the number of target genes (and target sites) per miRNA is highly nonuniform, ranging from zero for seven miRNAs to 268 for let-7b, with an average of 7.1 targets per miRNA. It is difficult to describe in detail, beyond the examples discussed in this text and beyond the annotation of target genes in Figure 3.2 and Table S3, which specific processes appear to be regulated by each miRNA or each set of co-expressed miRNAs. Groups of targets may reflect a reaction, a pathway, or a functional class (see Results). Although all miRNA–target pairs are subject to the condition of synchrony of expression, it is likely that typically one miRNA regulates the translation of a number of target messages and that, in some cases, the target genes as a group are involved in a particular cellular process. This was already known for the case of lin-4 (Ambros 2003).

The number of miRNA target sites per gene is also nonuniform, with a mean of 2.4. Although we do list target genes with single miRNA sites, there is increasing evidence that, in general, two or more sites are needed in the context of repression of translation. Although the details of these distributions (see Figure 3.2 and Table S3) depend on technical details, such as uniform cutoff for all miRNAs and evaluation in terms of a particular, imperfect scoring system, the general features of the distributions (see Figure 3.5) may be generally valid.

We conclude that multiplicity of targets and cooperative signal integration on target genes are key features of the control of translation by miRNAs. Neither multiplicity nor cooperativity is a novel feature in the regulation of gene expression. Indeed, regulation by transcription factors appears to be characterized, at least in eukaryotes, by analogous one-to-many and many-to-one relations between regulating factor and regulated genes (Kadonaga 2004). We are, of course, aware that the control cycles and feedback loops involving miRNAs cannot be adequately described without more detailed knowledge of the control of transcription of miRNA genes, about which little is known at present.

Although the predicted targets are subject to error (see estimate of false positives) and the prediction rules in need of improvement, several general principles of gene regulation by miRNAs are emerging. (1) Except in cases where a highly complementary match causes cleavage of the target message, miRNAs appear to act cooperatively, requiring two or more target sites per message, for either one or several different miRNAs. (2) Most miRNAs are involved in the translational regulation of several target genes, which in some cases are grouped into functional categories. (3) miRNAs carried in the context of RNPs appear to be sequence-specific adaptors guiding RNPs to particular target sequences. miRNA regulation of cellular messages may therefore range from a switch-like behavior (e.g., cleavage of mRNA message) to a subtle modulation of protein dosage in a cell through low-level translational repression (Bartel and Chen 2004).

These aspects of miRNA regulation complicate the design of experiments aiming at testing target predictions, or, more generally, at discovering biologically meaningful targets. Straightforward experiments that test one target site for one miRNA on one UTR will not be able to disentangle the effects of multiplicity or cooperativity. Tests for multiple sites on one UTR for one miRNA capture aspects of cooperativity (Doench and Sharp 2004), but still do not capture signal integration by diverse miRNAs. The most complicated situation is one in which multiple miRNAs affect multiple genes in combinatorial fashion, with fine-tuning depending on the state of the cell. We look forward to the results of ingenious experiments designed to deal with the complexity of miRNA regulation.

The results of this genome-wide prediction for mammals and fish are meant to be a guide to experiments that will in time elucidate the genetic control network of regulators of transcription, translation/maturation, and degradation of gene products, including miRNAs.

3.11.4   Implications for mechanisms of microRNA action

The role of a few animal miRNAs as posttranscriptional regulators of gene expression and, in particular, as inhibitors of translation is well established. However, the molecular mechanism of action is not well understood. Posttranscriptional control of protein levels can be achieved, for example, by cleaving the mRNA, by preventing RNP transport to ribosomes, by stalling or otherwise inhibiting translation on ribosomes, or by facilitating the formation of protein complexes near ribosomes that degrade nascent polypeptide chains. What do our results imply regarding the mechanism of action?

In analogy to plant miRNAs that have near perfect sequence complementarity and facilitate mRNA degradation, our predicted targets with near perfect complementarity between miRNA and mRNA plausibly are involved in mRNA cleavage (e.g., miR-196 and miR-138; see Results). Most of these would involve single target sites. In the case of Hox-B8, cleavage has been experimentally shown in mammalian cells (Yekta et al. 2004). We estimate that fewer than 5% of miRNA targets are cleaved as a result of miRNA binding.

Multiple target sites of lesser complementarity are consistent with RNP formation leading to translational inhibition, not mRNA degradation. Although we did predict single miRNA target sites for some genes, most target genes have multiple sites, indicating that cooperative binding (Doench and Sharp 2004) may be essential for formation of inhibitory RNP complexes.

An interesting and somewhat paradoxical feature is seen with mRNAs bound by FMRP, some of which increased and some of which are decreased in polysome fractions in FMRP knock-out mice (Brown, Jin et al. 2001)We see no bias in which of these two sets is most enhanced as predicted miRNA targets. This ambiguity not only raises questions about details of FMRP regulation but also raises the possibility that miRNA targets may not always be translationally repressed and may instead be translationally enhanced.

3.11.5   Improvement of prediction rules and algorithms

Current methods for predicting miRNA targets rely on conservation filters to reduce noise. Although the miRNA–mRNA pairings of experimentally validated targets were carefully used to define prediction rules, the information content in sequence match scores and free energy estimates of RNA duplex formation appears to be low. What is missing? Perhaps the fine details of experimentally proven target site matches are incorrect, although in some experiments mismatches and insertions have been tested. More plausibly, the rules do not yet capture additional functionally relevant interactions of miRNAs, such as in maturation and transport. Such additional interactions remain to be described in molecular detail, such as interactions with the small RNA processing machinery

(Drosha and Dicer) and with the components of RNPs (AGO and FMRP). A first step in this direction is the very recent analysis of the crystal structure of a PAZ domain of a human Argonaute protein, eIF2c1, complexed with a 9-mer RNA oligonucleotide in dimer configuration, which may represent three-dimensional interactions for the 3′ end of a miRNA (and siRNA) complexed, e.g., with Dicer or AGO (Ma, Ye et al. 2004). In this structure, each PAZ domain makes close binding contact with nine nucleotides of a single-stranded RNA. The two 3′ terminal nucleotides bind in a pocket through RNA backbone and other contacts. The remaining seven nucleotides bind PAZ through a series of backbone contacts such that nucleotides 3 to 9 are in an RNA helical conformation with bases exposed for base pairing to the second single-stranded RNA. If a 20–21-nt single-stranded RNA is bound to a PAZ domain in the same fashion, the 5′ end would be free for other interactions, such as binding to another protein domain in the RISC or base-pairing to mRNA. The conformational entropy that results when the 3′ end binds to PAZ, because the RNA helix is pre-formed, is consistent with weaker base pairing between miRNA and mRNA at the 3′ end of the miRNA, and stronger base pairing at the 5′ end. The dimeric structure of the PAZ domain (Ma et al. 2004) also raises the tantalizing possibility of cooperative binding of a dimer of two miRNA–PAZ combinations to two target sites on one or more mRNAs. In such an arrangement, seven residues at the 3′ ends of the two miRNAs (residues 3–9, but not the terminal two nucleotides) are paired in antiparallel fashion, with near perfect complementary pairing.

As more details of molecular contacts become available, prediction rules will evolve and improve in accuracy. The following elements are worth considering in the next generation of target prediction rules: (1) details of strand bias as deduced from siRNA experiments (Khvorova, Reynolds et al. 2003) (2) contribution of sequences outside of the mRNA target sites, (3) refinement of position-dependent rules, including different gap penalties for the mRNA and the miRNA, (4) energetics of miRNA–protein binding, starting with PAZ domain interaction, and (5) translation of systematic mutational profiling experiments into scoring rules. This will generate a more quantitative approach and enable machine learning methods to be more successful in microrNA and siRNA targeting overall.

3.11.6 Scope of microRNA regulation

The 160 mammalian microRNAs studied here have more than 8000 human targets, using 100 score threshold and 90% site conservation in rodents. Before the conservation filter there more than 90% of the aligned UTRs contained at least one non-conserved microRNA site at the same threshold, 18,000 genes. This is an order more than before conservation in rodents. Using the cutoffs described above, we predicted 5680 genes as targets with two or more microRNA target sites in their 3' UTRs conserved in mammals at 90% target site conservation (see Supplementary Tables 4 and 5). Thus, our prediction implies that *at least* 20% of protein-coding genes are under the regulation of these microRNA. In addition, we predicted another 2,320 genes with only one target site, but the false-positive rate for these is significantly higher, Figure 4.2. Some of the genes with single sites may contain additional sites that we cannot detect for a number of reasons, including truncated UTRs

4   Chapter Four – Competition and Saturation in microRNA regulation

4.1   Synopsis

Transfection of small RNAs (such as small interfering RNAs (siRNAs) and microR-NAs (miRNAs)) into cells typically lowers expression of many genes. Unexpectedly, increased expression of genes also occurs. We investigated whether this upregulation results from a saturation effect—that is, competition among the transfected si/miRNAs and the endogenous pool of miRNAs for the intracellular machinery that processes small RNAs. To test this hypothesis, we analyzed genome-wide transcript responses from more than 150 published transfection experiments in 7 different cell types. We show that targets of endogenous miRNAs are expressed at significantly higher levels after transfection, consistent with impaired effectiveness of endogenous miRNA repression. This effect exhibited concentration and temporal dependence. Strikingly, the profile of endogenous miRNAs can be largely inferred by correlating miRNA sites with gene expression changes after transfections. The competition and saturation effects have practical implications for miRNA target prediction, the design of siRNA and short hairpin RNA (shRNA) genomic screens and siRNA therapeutics.

All supplementary material referred to in this Chapter can be accessed either from the Nature Biotechnology website or http://cbio.mskcc.org/saturation/

4.2   Background

Thousands of microRNAs (21-23 nucleotide single-stranded RNAs) have been identified in animals over the past seven years(Ambros 2004; Ruvkun 2008). Research on miRNAs has focused on their biochemical processing and mechanism of action (Filipowicz, Bhattacharyya et al. 2008), the scope of their regulatory programs and their differential expression profiles in development and disease (He and Hannon 2004). Furthermore, various si/miRNA constructs are widely used in functional genomics, and microRNA cellular/tissue profiles are measured in medical diagnostics (Tam 2008). Finally, si/miRNAs (and their inhibitors) are in clinical trials for use as medical therapeutics (Leachman, Hickerson et al. 2008; Rossi, Zamore et al. 2008).

Contrary to expectations, however, some genes are strongly upregulated in si/miRNA transfections (Supplementary Fig. 1). Moreover, despite encouraging successes in using si/miRNAs in functional genomics and therapeutics, various unexpected effects have been reported, including a non-specific immune response (Robbins, Judge et al. 2008) and saturation of components of the sh/miRNA nuclear export machinery(Grimm, Streetz et al. 2006; McBride, Boudreau et al. 2008; Stewart, Li et al. 2008), such as *exportin-5*. It has been suggested that saturation-related effects can be avoided by using siRNAs(McBride, Boudreau et al. 2008)(since they do not rely on the nuclear export machinery) rather than shRNAs. Indeed, a recent prominent report specifically claimed that effective siRNAs targeting *APOB* and *F7* do not interfere with endogenous miRNA function(John, Constien et al. 2007). However, use of siRNAs have not been problem-free, as scrambled siRNAs have been shown to cause dose-dependent upregulation of a target gene, *SREBF1*, in three different cell types(Castanotto, Sakurai et al. 2007; Vankoningsloo, de Longueville et al. 2008)13, and an elegant report on combinatorial delivery of siRNAs in HEK 293 cell lines demonstrated competition for RISC (RNA-induced silencing complex) machinery(Castanotto, Sakurai et al. 2007).

Here, we investigate the hypothesis that the unexplained si/miRNA-induced gene upregulation is due, at least partly, to a loss of function of endogenous miRNAs, as modeled in Figure 4.1 and supported by previous reports (Castanotto, Sakurai et al. 2007). In this model, transfected small RNAs compete with endogenous miRNAs for the RISC complex or other machinery further downstream than *exportin-5* in the miRNA pathway, such as *Argonaute* proteins or *TRBP* (Wargelius, Ellingsen et al. 1999; Oates, Bruce et al. 2000; Gruber, Lampe et al. 2005; Castanotto, Sakurai et al. 2007). Loss of available RISC through competition would be expected to relieve repression of the targets of endogenous miRNAs—an effect that may be observed as upregulation of target mRNAs and corresponding proteins. We reasoned that we should be able to detect this effect in gene expression profiles measured after si/miRNA perturbations. We also reasoned that this effect may be observable in the dose response and temporal dynamics of the misregulated off-target gene(Jackson, Bartz et al. 2003; Burchard, Jackson et al. 2009)

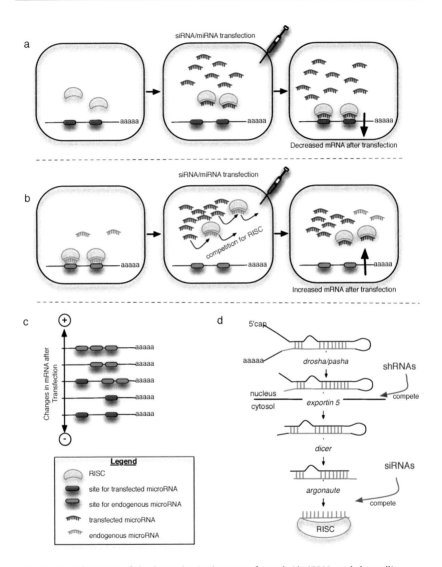

Figure 4.1 Schematic of the hypothesis that transfected si/miRNA and the cell's endogenous miRNAs compete for RISC machinery. *(a) Genes with sites (red) for the transfected small RNA (red) are downregulated after transfection. Genes with sites for endogenous miRNAs (green) may be upregulated after small RNA transfections.*
(b) Biogenesis of miRNAs. sh- and siRNAs enter the miRNA processing pathway at different points.

4.3   Methods

4.3.1   mRNA and protein experimental datasets

We collected data from four types of experiments: (i) transfection of a miRNA followed by mRNA profiling using microarrays (Lim, Lau et al. 2005; Jackson, Burchard et al. 2006; Jackson, Burchard et al. 2006; Wang and Wang 2006; He, He et al. 2007; Linsley, Schelter et al. 2007; Selbach, Schwanhausser et al. 2008), see Table X ; (ii) transfection of an siRNA followed by mRNA profiling(Jackson, Bartz et al. 2003; Jackson, Burchard et al. 2006; Burchard, Jackson et al. 2009); (iii) inhibition of miRNA followed by mRNA profiling (Krutzfeldt, Rajewsky et al. 2005; Elmen, Lindow et al. 2008) ; and (iv) transfection of miRNA followed by protein profiling using mass spectrometry (Baek, Villen et al. 2008; Selbach, Schwanhausser et al. 2008). These four types of datasets of 150 experiments encompass 7 different cell types, 20 different miRNAs, and 40 different siRNAs (Supplementary Table X). The synthetic transfected miRNAs are all commercially available siRNA/miRNA mimics (Dharmacon, Inc.). Sequences of mimics can be found in the respective references. When possible, we used normalized micro-array expression data as provided with the original publications. In all other cases, we used the "affy" package in the "R" software package to perform RMA normalization of microarray probe-level data. For statistical analysis over mul-tiple mRNA microarray profiling experiments, each experiment was indepen-dently centered using the mean $log_2(expression\ change)$ of genes lacking con-served endogenous or exogenous sites and normalized to have unit variance in $log_2(expression\ change)$ across all genes. This normalization results in a modified Z-transformation of the data, where genes with no exogenous or endogenous sites have mean 0. For the transfection experiments followed by mass spectrom-etry, we used normalized protein expression levels as provided by the authors of the original publication, Supplementary Fig. 2.

Transfected microRNA/siRNA	Cell type
miR-124, miR-1, miR-373, miR-124mut5-6, miR-124mut9-10, chimiR-1-124, chimiR-124-1	HeLa
miR-106b, miR-200a/b, miR-141 miR-16, miR-15a/b, miR-103, miR-107, miR-192, miR-215, miR-17-5p, miR-20, let-7c, miR-195	HeLa, HCT116 HCT116 Dicer$^{-/-}$
miR-7, miR-9, miR-122a, miR-128a, miR-132,miR-133a, miR-142, miR-148, miR-181a	HeLa
miR-34a/b/c	HeLa, A549, TOV21G, HCT116 Dicer$^{-/-}$
miR-1, miR-155, let-7b, miR-30	HeLa
miR-181a, miR-124, miR-1	HeLa
miR-34a	HeLa
miR-124	HepG2
igfr-1-16, mapk14-1-8, mapk14-1 (1,2,4,6,12,24,48,72,96) hours, mapk14-1 (.16,.8,4,20,100) nM, mapk14-1 (pos 4,5,15) mismatch	HeLa
mphosph1-2692, pik3ca-2692, prkce-1295, vhl-2651, vhl-2652, sos1-1582	HeLa
pik3cb-6338, plk-1319, plk-772, pik3cb-6340	HeLa
siApoB-Hs1/Hs2/Hs3/Hs4	Huh7

Table 4.1

4.3.2   Target prediction

We conducted four different types of miRNA target site searches using miRNA sequences grouped into families, and 3' UTR alignment of 5 species. miRNAs were grouped into families as defined by identical nucleotides in positions 2-8. We searched for target sites for miRNA families in 3'UTRs using four different types of seed matches: (i) 6-mers (position 2-7 and 3-8), (ii) 7-mers (position 2-8), (iii) 7-mer positions 2-7 m1A (the first nucleotide an A in the mRNA) and (iv) 8-mers (position 1-8). 7-mer positions 2-8 were selected for analysis since this choice gave the most significant p-values for downregulation of targets with sites for the transfected si/miRNA based as compared to baseline genes based on a one-sided KS statistic (set X versus set B, as described below).

For target matches, we considered both non-conserved and conserved targets in human 3'UTRs. 3'UTR sequences for human (hg18), mouse (mm8), rat (rn4), dog (canFam2), and chicken (galGal2) were derived from RefSeq and the UCSC genome browser (http://hgdownload.cse.ucsc.edu/goldenPath/hg18/multiz17way/). We used multiple genome alignments across the 5 species as derived by multiZ. The RefSeq annotation with the longest UTR mapped to a single gene was always used. To establish a conservation filter, we required that the 7-mer target site in human be present in at least three of the other four species, i.e. exact matching in a 7 nucleotide window of the alignment in at least 3 other species, to be flagged as conserved. Restricting to conserved sites led to more significant p-values for downregulation of targets with exogenous sites as compared to baseline genes (one-sided KS statistic, set X versus set B, as defined below). We chose these stringent requirements so that our prediction method would be conservative and err on the side of under-prediction rather than over-prediction. However, we acknowledge that there are indeed functional siRNA and miRNA target sites that have mismatches, G:U wobbles in the 5' end and are not conserved (see for example work of the Hobert and Slack groups) (Vella, Reinert et al. 2004; Didiano and Hobert 2006).

4.3.3   Endogenous miRNA expression.

We used endogenous miRNA profiles from the Landgraf *et al.* (Landgraf, Rusu et al. 2007) compendium for HeLa, A549, HepG2 and TOV21G, which provide relative miRNA expression levels from cloning and sequencing small RNA libraries. We used miRNA profiles from the Cummins *et al* (Cummins, He et al. 2006) cloning and sequencing data for HCT116 and HCT116 Dicer$^{-/-}$. For consistency across cell types, we took the top 10 miRNAs with highest expression levels (clone counts), which corresponds to at least 75% of the miRNA content in each cell type, to be the set of endogenous miRNAs in our statistical analysis.

4.3.4 Kolomogorov-Smirnov statistics.

To compare the expression changes for two gene sets, we compared their distributions of Z-transformed *log$_2$(expression change)* using a one-sided Kolmogorov-Smirnov (KS) statistic, which assesses whether the distribution of expression changes for one set is significantly shifted downwards (downregulated) compared to the distribution for the other set. We chose the KS statistic to apply a uniform treatment of data despite the heterogeneity of the transfection experiments, which involve different cell types, different numbers of target genes with sites for the transfected si/miRNA, and different apparent transfection efficiencies. The KS statistic has the advantages that (i) it is non-parametric and hence does not rely on distributional assumptions about expression changes; (ii) it does not rely on arbitrary thresholds; and (iii) it measures significant shifts between the entire distributions rather than just comparing the tails. The KS statistic computes the maximum difference in value of the empirical cumulative distribution functions (cdfs):

$$\sup_x \left(F_1(x) - F_2(x) \right) \quad \text{where} \quad F_j(x) = \frac{1}{n_j} \sum_{i=1}^{n_j} I_{X_i \leq x}$$

is the empirical cdf for gene set $j = 1, 2$, based on n_j (Z-transformed) *log$_2$ (expression change)* values. We used the Matlab function kstest2 to calculate the KS test statistic and asymptotic *p*-value. Full KS test results are provided in Supplementary Table 2.

4.3.5 Notation

We use the following notation to describe sets of genes based on the number of sites for exogenous and endogenous miRNAs in their 3¢UTRs:

Non-conserved sites: sets with subscript $_{NC}$ denote non-conserved sites have been used; subscript $_{NC}$2 denotes 2 or more non-conserved sites

Endogenous sites: sites for endogenous miRNAs, i.e., miRNAs expressed in the cell

Exogenous sites: sites for exogenous si/miRNAs, i.e., small RNAs introduced into the cell

X ("*eXogenous*"): set of genes containing at least one site for the exogenous (transfected) si/miRNA

D ("*enDogenous*"): set of genes containing at least one site for a miRNA endogenously expression in the cell type

B ("*Baseline*"): set of genes containing neither exogenous nor endogenous sites

D-X: set of genes containing at least one endogenous site and no exogenous sites

X∩D: set of genes containing at least one exogenous site and at least one endogenous site

X-D: set of genes containing at least one exogenous site and no endogenous sites

D^2: set of genes containing 2 or more endogenous sites

X∩D²: set of genes containing at least one exogenous site and at least 2 endogenous sites

4.3.6   Regression analysis to model expression

We performed multiple linear regression to fit a linear model expressing the Z-transformed *log₂(expression change)*, denoted as *y*, in terms of the number of a gene's exogenous and endogenous target sites, denoted as n_X and n_D, respectively:

$$y = c_X n_X + c_D n_D + b.$$

We use the Matlab *regress* function to fit the model and assess the significance of the fit as measured by the R^2 statistic. We used the *F* statistic, also computed by the *regress* function, to assess whether the linear model with 2 independent variables, n_X and n_D, significantly improves the fit over the simpler model:

$y = c_X n_X + b$ given the number of sites for exogenous si/miRNAs *a priori*. All *p*-values from the *F* statistic across experiments are reported in Supplementary Table X.

4.3.7   Forward stepwise regression analysis

As an extension to the linear model with 2 independent variables, we performed forward stepwise regression to fit the number of target sites for each of the (162) miRNA families to the Z-transformed *log₂(expression change)* data. Starting again with the simpler model, $y = c_X n_X + b$, we incrementally added the number of target sites for the miRNA seed family with highest F statistic to the model. The procedure was continued until the p-value from the F-statistic for the best remaining seed family failed to satisfy a significance threshold of $P <$ 0.05. The final model can be viewed as a linear combination of the number exogenous target sites and the additive contribution of other miRNAs represented by their number of target sites n_i:

$$y = c_X n_X + \sum_i c_i n_i + b$$

Since we did not enforce a stringent significance criterion for including miRNA sites in the model, we do not expect every miRNA added to the model to be correct; however, miRNAs added consistently across different transfections experiments are likely to be significant. We repeated the forward stepwise regression for multiple experiments in HeLa and HCT116 Dicer$^{-/-}$ cells and computed the frequency of the most statistically significant additive factors with positive regression coefficient in the model for each cell type; we reported the 10 most frequent of these miRNAs. All p-values from the F statistic across experiments are reported in Supplementary Table 4.

4.3.8   Cell cycle and cancer genes analysis

A list of expertly annotated genes for which mutations (both germline and somatic) have been causally implicated in cancer was obtained from the Cancer Genome Project (Cancer Gene Census catalog version 2008.12.16, http://www.sanger.ac.uk/genetics/CGP/Census) (Stratton, Campbell et al. 2009). A list of genes that have consistently showed a periodic expression pattern during the cell cycle in several mRNA microarray studies was obtained from the Cyclebase database (Gauthier, Larsen et al. 2008) . From these lists, we could match 312 and 651 genes to the mRNA datasets collected in this work, respectively. The gene sets were designated "oncogenes" and "cell cycle genes", respectively. To investigate if oncogenes or cell cycle genes were enriched for miRNA targets in Hela cells compared to all genes we used Fisher's exact tests.

4.4   Results

4.4.1   Targets of endogenous miRNAs are upregulated after miRNA transfection

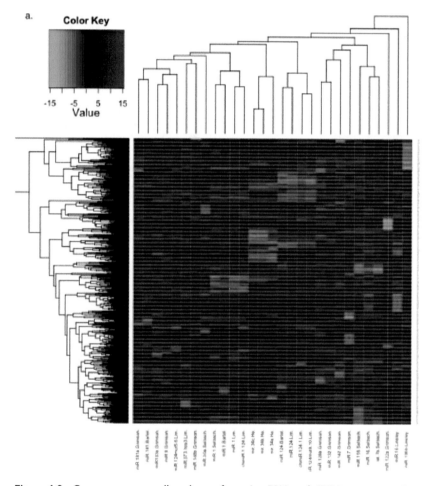

Figure 4.2a Genes go up as well as down after microRNA and siRNA overexpression

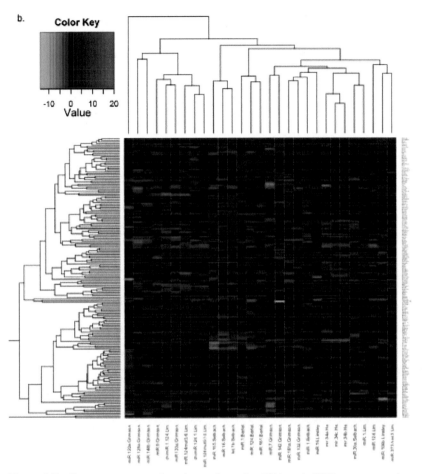

Figure 4.2b Genes go up as well as down after microRNA and siRNA overexpression

First we pooled all the data from experiments in our panel done in HeLa cells. We confirmed that a significant number of genes up-regulated as well as those you would expect to be down-regulated, Figure 4.2 To test the hypothesis presented in Figure 4.1, we assembled data from published experiments in which small RNAs were transfected into cells in culture which were then assayed using mRNA profiling or protein mass spectrometry (Methods and Supplementary Fig. 2). In total, we gathered data from 151 experiments from 7 different cell types, involving 29 different miRNAs (as well as 2 mutant and 2 chimeric miRNAs) and

42 unique siRNAs (Table 1 and Supplementary Tables 1 and 2). Strikingly, a large number of genes are upregulated in the si/miRNA experiments, rather than downregulated as would be expected (Supplementary Fig. 1).

To investigate whether predicted targets of cellular (endogenous) miRNAs respond to transfected si/miRNAs, we assessed global expression changes following si/miRNA transfection or miRNA inhibition (Methods). Briefly, we used available miRNA expression profiles to define the 10 most highly expressed endogenous miRNAs in each cell type, which together make up 70–80% of the measured cellular miRNA content ,(He, He et al. 2007; Landgraf, Rusu et al. 2007) (Supplementary Fig. 3). We then identified genes with predicted sites targeted by these 10 'endogenous' miRNAs (hereafter abbreviated D), genes with predicted sites for the 'exogenous' transfected si/miRNA (hereafter abbreviated X), and a 'baseline' set of genes with neither endogenous nor exogenous sites (hereafter abbreviated B). Differences in global expression changes between gene sets following perturbation were assessed for statistical significance by a one-sided Kolmogorov-Smirnov (KS) test (Methods).

The following example illustrates our findings: when miR-124 is transfected into HeLa cells, genes with sites for HeLa-expressed (endogenous) miRNAs and without miR-124 sites (that is, genes in set D but not in set X, hereafter abbreviated as D-X) are significantly upregulated compared to the baseline set ($P < 7.5e\text{-}34$; Fig. 2a green line, and Supplementary Table 2). The magnitude of upregulation was even greater when we limited our analysis to genes with at least 2 endogenous sites and no sites for the transfected miRNA ($P < 2.2e\text{-}24$; Figure 4.3, blue line). Overall, we observed effects of this kind in 89% of the miRNA transfection experiments tested (using significance threshold of $P < .05$, n=61) (Supplementary Table 2).

To further investigate whether this is a general effect, we pooled all of the HeLa transfection experiments and repeated the analysis. We found that the set D-X was significantly upregulated in the pooled HeLa data ($P < 10^{\text{-}100}$; Fig. 2b). The same was also true for pooled data in A549, HCT116, HCT116 Dicer$^{/\text{-}}$ and Tov21G cells, with $P < 10^{\text{-}10}$ for all cell types (Supplementary Table 2). Interestingly, in experiments in which an endogenous miRNA was over expressed, the targets of other endogenously expressed miRNAs were also upregulated. For example, when HeLa cells were transfected with the endogenously expressed miRNAs miR-16 or let-7b, the set D-X was upregulated compared to the baseline set ($P < 5.6e\text{-}19$, $P < 6.1e\text{-}12$ respectively; Supplementary Table 2, Supplementary Fig. 4).

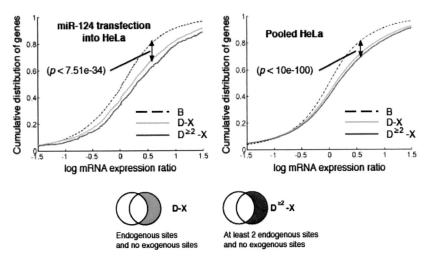

Figure 4.2 Genes with predicted target sites for endogenous miRNAs are significantly dysregulated after si/miRNA transfections. *P-values shown are calculated by one-sided KS test as described in Methods. See bottom of figure for notation used to abbreviate gene sets tested for significant expression changes (D-X, D^{≥2}-X,) relative to a baseline set of genes (B). (a) miR-124 transfection results in up-shift in gene expression for D-X. (b) Pooled data from 15 miRNA transfections into HeLa cells (miR-373, miR-124, miR-148b, miR-106b, miR-124, mut9-10, miR-1, miR-181, chimiR-124-1, chimiR-1-124, miR-16, mir-34a, mir-34b, miR-128a and miR-9)*

As a positive control, to verify that transfected small RNAs affected their predicted targets, we compared changes in expression of the targets of the exogenous RNAs to the baseline gene set, both in individual transfection experiments and in sets of experiments grouped by cell type (Table 1). As expected, we found that expression of the target mRNAs was significantly downshifted compared to the baseline set (Supplementary Table 2).

We also investigated whether the effect of miRNA transfection extends to protein levels by analyzing HeLa cell data from mass spectrometry experiments following miRNA transfection24. Significant changes in protein levels were observed (Supplementary Table 2).

In particular, protein levels of target genes with sites for endogenous miRNAs and no sites for exogenous miRNAs (genes D-X) were upregulated when compared to baseline genes ($P < 1.3e-9$, pooled data). For example, when HeLa cells were transfected with the endogenously expressed let-7b, protein levels of

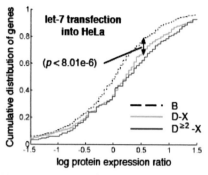

Figure 4.3 Protein changes after microRNA overexpression

genes with other endogenous target sites increased significantly relative to the baseline ($P < 8e\text{-}6$; Figure 4.3, green line). Taken together, we find that global perturbations that follow miRNA transfection are consistent with a 'competition effect' in which the transfected miRNA competes with endogenous miRNAs for cellular protein machinery.

4.4.2   siRNA transfections display the same effect as miRNA transfections

To investigate whether siRNAs perturb the transcriptome similarly to miRNAs, we analyzed gene expression profiles of 42 siRNA transfections in HeLa cells[19,25,26]. As a representative example of our results we found that targets of endogenous miRNAs were significantly upregulated after transfection of a siRNA that targets *MAPK14* (Figure 4.4).

Upregulation of targets of let-7 and miR-15, two miRNAs highly expressed in HeLa cells, was especially significant (Supplementary Table 3). Pooling the data from these siRNA experiments, we see a significant upward shift in the expression of genes with endogenous sites only relative to the baseline gene set ($P < 10\text{-}100$; Supplementary Table 2). Five different siRNAs designed to target *VHL*, *PRKCE*, *MPHOSPH1*, *SOS1* and *PIK3CA26*, respectively. These siRNAs showed upregulation of similar sets of genes including *CCND1*, *DUSP4*, *DUSP5* and *ATF3* (Supplementary Table 3), despite the fact that their direct targets were different. As each of these upregulated genes contains at least one site for an endogenous miRNA,

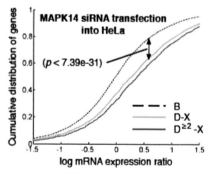

Figure 4.4 siRNAs also cause upregulation of predicted endogenous microRNA targets

this observation is consistent with upregulation as a consequence of the siRNA transfection, independent of the specific siRNA sequence.

4.4.3 Attenuated knockdown of si/miRNA targets containing endogenous sites

To investigate whether the 'competition effect' might attenuate the strength of si/miRNA knockdown, we analyzed the expression of genes directly targeted by the transfected si/miRNA. Specifically, we partitioned the set of genes with sites for transfected miRNAs (set X) into two subsets – those without endogenous sites (labeled X-D) and those with endogenous sites (labeled XÇD). As a representative example of our results after transfection of miR-16 into HeLa cells, predicted miR-16 target genes without endogenous sites (that is, X-D) were downregulated significantly more than targets with endogenous sites (that is, XÇD) (P < 1.2e-3, Figure 4.5).

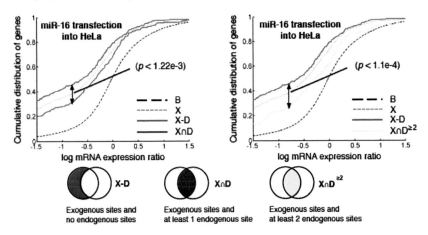

Figure 4.5 Attenuated knockdown of genes with microRNA sites after RNAi
P-values shown are calculated by one-sided KS test as described in Methods.). (a) miR-124 transfection results in up-shift in gene expression for D-X. (b) Pooled data from 15 miRNA transfections into HeLa cells (miR-373, miR-124, miR-148b, miR-106b, miR-124, mut9-10, miR-1, miR-181, chimiR-124-1, chimiR-1-124, miR-16, mir-34a, mir-34b, miR-128a and miR-9). (c) Protein expression changes after let-7b transfection. (d) mRNA expression changes after MAPK14-siRNA is transfected into HeLa cells showing upregulation of genes with sites for endogenous miRNAs (green line). (e, f) Genes that contain sites for both endogenous miRNA and transfected small RNAs are less downregulated than if they contain only sites for transfected small RNAs.

Limiting the analysis to genes with two or more sites for endogenous miRNAs resulted in an even more pronounced difference ($P < 1.1e-4$, Fig. 2f, yellow line). Pooling data across a panel of transfection experiments into HeLa cells gave a highly significant result ($P < 3.6e-13$, Supplementary Table 2). Taken together, these results suggest that competition with cellular machinery may attenuate the effectiveness of si/miRNA knockdown.

4.4.4   A quantitative model resolves the endogenous miRNA profile

To strengthen our analysis and predict the saturation effect on individual genes, we built a quantitative mathematical model of the change in gene expression after si/miRNA transfection. This model can be used to predict which genes are likely to be upregulated or downregulated (off-target effects) after si/miRNA transfections. Considering each transfection into HeLa cells independently, we first fit a simple linear regression model (Methods) to predict the change in expression of genes based on the number of exogenous sites (n_x) and the number of endogenous sites (n_D) in their 3′ UTR (Figure 4.6).

In a large majority of experiments, the endogenous count n_D was found to be a significant variable for explaining expression changes (84 out of 109 experiments satisfying $P < 0.05$ by F statistic, Supplementary Table 2). As expected, the regression coefficient for the endogenous count was always positive when significant, meaning that these sites correlate with upregulation, while the regression coefficient for the exogenous count was always negative. Figure 4.1b is a cartoon version of the expected effect on expression of a gene that contains different combinations of exogenous and endogenous sites.

We then refined the model to ask whether the presence of sites recognized by individual endogenous miRNAs could explain upregulation of targets in an experiment. More generally, we considered *all* human miRNA families as potential variables in the regression model and assessed whether sites of individual miRNAs accounted for expression changes in a transfection experiment. We ranked the importance of each miRNA by the number of experiments in which it was included in a forward stepwise regression model (Methods). Among the 10 most frequently included miRNAs, we identified 7 of the most highly expressed miRNAs in a HeLa and 4 of the most highly expressed in HCT116 Dicer[-/-] cells, using no prior knowledge of the miRNAs expressed in those cell types (Figure 4.6). The top ranked miRNAs retrieved by this analysis, let-7 and miR-21, are

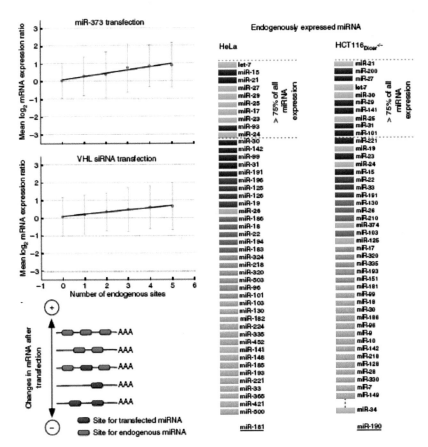

Figure 4.6 Quantitative model predicting expression change after transfection. *(a) Linear regression fitting mean log2(expression change) to number of endogenous sites in genes having no exogenous sites, in miR-373 and VHL-siRNA experiments. One s.d. from mean depicted witherror bars. (b) Cartoon of genes with different combinations of sites showing net effect on gene expression. (c) Endogenous miRNAs expressed in HeLa cells and HCT116 Dicer–/– (ranked in orderof endogenous expression, with black to gray depicting decreasing relative expression). Green rectangles indicate the miRNAs that occurred*
most often with positive regression coefficient in stepwise regression models for 420 HeLa cell and 16 HCT116 Dicer–/– transfection experiments (Methods). These miRNAs can be interpreted as a predicted endogenous profile.
Underlined miRNAs were predicted by regression analysis but are not known to be expressed endogenously.

thought to be the most highly expressed miRNA in HeLa and HCT116 Dicer[-/-] cells, respectively, therefore supporting a saturation model. Taken together, these results suggest that the endogenous miRNA profile in a cell can be largely determined from expression changes after transfection of small RNAs, which plausibly are due to competition for cellular resources.

4.4.5   The competition effect has a dose response

In a previous study19, siRNA dose response was investigated by transfecting a siRNA targeting *MAPK14* into HeLa cells in a range of 5 doses, from 0.16nM-100nM, followed by microarray profiling after 24 h. As described in the original publication, we observed that off-target genes (that is, genes other than *MAPK14* with sites in their 3¢ UTR for the siRNA) were affected in a dose-responsive manner that mimicked the dose response of the main target (*MAPK14*) and that these off-target effects were not titrated away at lower transfection concentrations. However, we also found that many genes with sites for endogenous miRNAs follow a pattern of upregulation that mirrors the downregulation of off-targets (Figure 4.7). More specifically, a fivefold change in siRNA dose from 4 nM to 20 nM produced a twofold change in mean gene expression of the most responsive upregulated genes and the most responsive downregulated genes. The change in expression of both the endogenous target and off-target sets

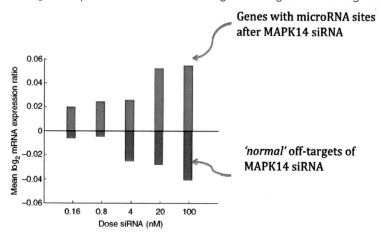

Figure 4.7 The competition effect shows a dose-response *Effect of dose of siRNA transfection on mean log2(expression change) of X_{NC} (gray) and 90th percentile of D^{zz} – X_{NC} (green)*

reaches near-maximal dose response at 20 nM. In summary, these siRNA satura-tion effects and off-target effects roughly scale with the dose response of the main target, at least for a significant fraction of genes in these sets, and we did not observe that they were titrated away at lower transfection concentrations.

4.4.6   Evidence for a transitory saturation effect

To measure the time dependence of the response of genes with sites for endog-enous miRNAs, we examined published data in which gene expression changes were monitored over a period of 96 h after transfection of a siRNA targeting *MAPK14*. Given our previous observations that genes putatively regulated by endogenous miRNAs were de-repressed after transfection, we expected that the temporal response of these genes would be similar to that of the intended siRNA target gene and the off-targets (that is, genes with non-conserved seed matches, X_{NC}), Figure 4.7. We compared the mRNA changes of the putative off-target genes to *MAPK14* mRNA itself. Although the off-target genes followed a temporal downregulation pattern similar to *MAPK14* in the first 48 h, the ex-pression level of the X_{NC} set of genes returned to near their original expression level by 92 h. In contrast the intended target *MAPK14* had a gradually increasing downregulatory effect, with a half maximal effect at ~12 h and a sustained effect from 24-96 h (Figure 4.9; light-green bar).

We investigated the dynamics of a set of genes with at least 2 non-conserved endogenous sites (90th percentile for expression change, pooling all time points, ~1000 genes), compared to a set of siRNA targeted genes (Methods).

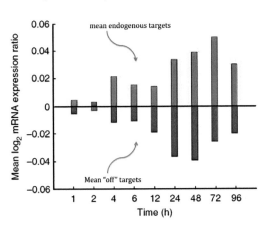

Figure 4.8 Competition effect shows temporal dynamics proportional in magnitude, but opposite in direction, to targeted effect. *Mean log2 (expression change) of XNC (gray) and 90th percentile of DZ2 – XNC (green) versus time.*

The genes in the endogenous set had maximal upregulation at 24-48 h with similar dynamics across the 92 h, consistent with being targets of endogenous miRNAs competing for components of the RISC, Figure 4.8. This similarity was also apparent in the expression patterns of the 6 most downregulated 'off-target genes' and the 6 most upregulated genes in set D_{NC}-X_{NC}, Figure 4.9.

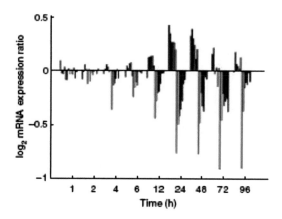

Figure 4.9 Competition effect for specific genes shows smae temporal dynamics as direct target gene. *Mean log_2(expression change) of X_{NC}(gray) and 90th percentile of $D_{NC}^32 - X_{NC}$ (green) versus time. (c) log_2(expression change) of putative endogenously regulated genes (SCML2, TNRC6, YOD1, CX3CL1, AKAP12 and PGM2L1) and a set of MAPK14-siRNA 'off-targets' (MAPK14 (light-green), MARK2, SLC35F3, HMGB3, FZD7, RPA2 and IER5L) over same time course as Figure 4.8 (genes displayed in this order, from left to right).*

The upregulated genes (*SCML2, TNRC6, YOD1, CX3CL1, AKAP12* and *PGM2L1*) each contain at least 4 sites for highly expressed endogenous miRNAs. The expression patterns of these genes are consistent with the model that they are targets of endogenous miRNAs competing for components of the RISC. Since *TNRC6* is associated with *AGO2* in RISC, this inadvertent TNRC6 upregualtion may in turn affect the function of all microRNAs expressed in the cell(Landthaler, Gaidatzis et al. 2008)

We also investigated published experiments that were designed to examine the off-target effects of four therapeutic siRNAs targeting the human coronary artery disease target gene, *APOB (Burchard, Jackson et al. 2009)*. We observed a significant saturation effect with all of the siRNAs ($P <$ 1e-8 at 6 h, Supplementary Table 2) and that this effect reached its maximum at 6 h. Taken together, these investigations of the temporal dynamics of small RNA gene regulation after transfection show that the upregulatory effect mirrors the expected downregulatory effect, supporting the proposed competition model.

(Supplementary Fig. 5b). Taken together, these observations raise the possibility that cell cycle and oncogenes may be particularly susceptible to the proposed saturation effect.

4.4.7   Overexpression of Argonaute shows no saturation effect

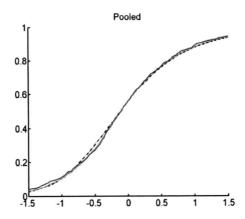

Pooled

Figure 4.10 microRNA overexpression with AGO2 overepxression. *No saturation effect is send as expected when Argonaute is overepxressed.*

4.4.8   miRNA inhibition may cause upregulation of other endogenous targets

Finally, we examined published data (Linsley, Schelter et al. 2007) that measured expression changes after miRNA inhibition by 'antagomirs', which are chemically modified single-stranded RNA analogues that inhibit a specific target miRNA, Figure 4.11. Treatment of cells with antagomirs to miR-16 and miR-106b significantly upregulated genes that contained only endogenous sites ($P <$ 5e-16 (D-X) and $P <$ 2e-30 (D'2-X)), including *SSR3*, *PLSCR4* and *PTRF* (Supplementary Table 3) Moreover, inhibition of miR-122 with locked nucleic acid (LNA) molecules also produced significant upregulation of genes with sites for other endogenous miRNAs when compared with a saline transfection (*P* < 2.5e-6). Elmen *et al.* observed dose-dependent accumulation of a shifted heteroduplex band, implying that the LNA-antimiR binds stably to the miRNA (Elmen, Lindow et al. 2008).

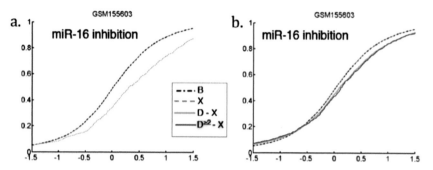

Figure 4.11 Genes with endogenous microRNA regulation are upregulated after microRNA inhibition. a. *CDF of change in genes expression of all genes (black and genes with sites for miR-16 (grey) expressed in HeLa cells* **b.***CDF of genes with sites for endogenous microRNAs (green =1, blue = more than 1)*

This finding is consistent with hypothesis that the heteroduplex of miR-122::antimiR prevents the availability of free RISC machinery (Supplementary Fig. 6), but clearly more experiments are needed to distinguish between the possible models and to assess the size of the inhibition effect on the function of endogenous miRNAs.

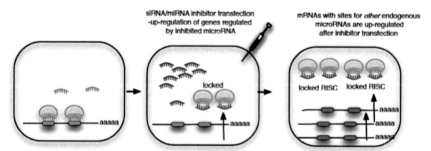

Figure 4.12 Other endogenous microRNA target genes are up-regulated after inhibition of one particular endogenous microRNA.. *Schematic of hypothesis consistent with these inhibition results*

4.4.9   microRNAs regulate cancer associated genes and these are dysregualted during small RNA transfections

Dysregulation of endogenous miRNAs is known to contribute to tumorigenesis (Ventura and Jacks 2009), so we tested whether cancer associated genes were disproportionately perturbed after small RNA transfections. To do this we used the cancer census set of genes ~200 genes originally published in 2004 (Futreal, Coin et al. 2004; Forbes, Bhamra et al. 2008) and now expanded to over 400, www.sanger.ac.uk/genetics/CGP/Census/.

A significant number of these cancer associated genes Figure 4.13 were consistently upregulated across the transfection experiments. For instance, known miRNA targets, including the oncogenes *HMGA2*, *CCND1* (Linsley, Schelter et al. 2007; Mayr, Hemann et al. 2007) and *DUSP2*, are upregulated after many different independent HeLa transfection experiments, including siRNA transfections. We also find that cell cycle genes are significantly enriched in endogenous miRNA target sites compared to other genes expressed in HeLa. Figure 4.14 shows examples of genes which are upregulated after microRNA transfections, when they have sites for endogenous microRNAs. Strikingly, 4 of these are also detected upregulated at the protein level by mass spectrometry in HeLa cells after the microRNA overexpression.

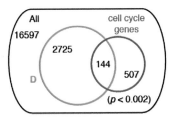

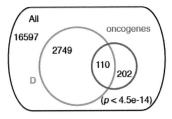

Figure 4.13 Enrichment of cell cycle genes and oncogenes in upregulated genes. *Venn diagrams showing overlap between genes with endogenous miRNA target sites in HeLa cells and cell cycle genes or oncogenes; the significance of the overlap is assessed by Fisher's exact test.*

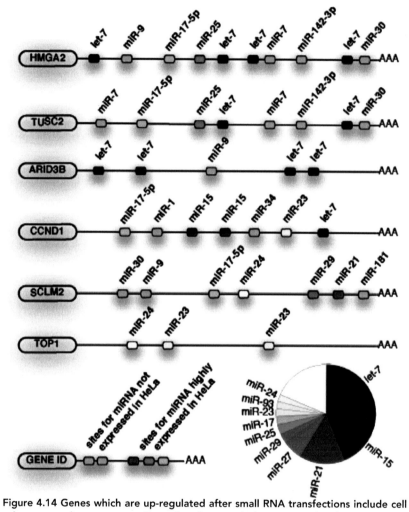

Figure 4.14 Genes which are up-regulated after small RNA transfections include cell cyle genes and oncogenes. a. Examples of genes which are up-regulated in experiments when they contain sites for endogenous mciroRNAs but not the transfected microRNA. Colours of endogenous sites match expression profile shown; grey sites represent transfected microRNA sites. Only sites for endogenous or transfected microRNAs are shown. b. Representation of relative expression of top 10 families of microRNAs in HeLa cells.

4.5   Summary

We have shown that transfecting small RNAs affects the expression of genes predicted to be under endogenous miRNA regulation. This effect is observable at both the mRNA and protein levels. Moreover, this effect is observable in experiments that use siRNAs that target particular genes, and in experiments that use miRNA mimics and miRNA inhibitors designed to test the biological effects of miRNAs. Using a quantitative approach, we built a regression model that can identify many of the endogenous miRNAs expressed in a cell type simply from the changes in gene expression following small RNA transfections. The purpose of this approach is not to infer miRNA profiles *per se* but to provide independent evidence of the indirect perturbation of miRNA function. Finally, we used published data to show that the temporal dynamics and dose response of genes affected by the proposed competition effect follow the same patterns as those of the genes directly targeted by the transfection.

The most plausible model for these observations is saturation of the RISC complex (or other necessary small RNA processing or transport machinery) and competition between the transfected small RNA and endogenous miRNAs for binding ,Figure 4.1. However, other models that may be consistent with the observed effect cannot be ruled out by our analyses. While the precise mechanism of this competition effect remains to be established, the statistical significance of the observed shifts in transcript levels is clear, and the results of these analyses support the thesis that small RNA transfections unexpectedly and unintentionally (from the point of view of the investigators) disturb gene regulation by endogenous miRNA.

Our results have potentially important practical consequences for the use of siRNAs and shRNAs in functional genomics experiments. While it is already known that siRNAs can produce unwanted off-target effects, such as unintended downregulation of mRNAs via a partial sequence match between the siRNA and target, the effects observed here are distinct and involve the de-repression of miRNA-regulated genes.

Our findings also have consequences for the development of miRNA target prediction methods. Since measuring mRNA expression changes after si/miRNA perturbations is a standard way to validate miRNA target prediction methods (Lim, Lau et al. 2005; Grimson, Farh et al. 2007; Elmen, Lindow et al. 2008) one should take the saturation effect into consideration. Despite concerted efforts, bioinformatic si/miRNA target prediction methods still significantly over-predict the number of targets by at least 7 fold(Grimson, Farh et al. 2007; Selbach, Schwanhausser et al. 2008; Bartel 2009). Elegant work showing the dynamic (condition and cell-type dependent) regulation of UTR lengths (Sandberg, Neilson et al. 2008) may explain some of these false positives, since shortening of UTRs may lead to loss of target sites, but is unlikely to explain all. The proposed competition effect may offer an explanation for false positive target prediction in cases where UTRs have target sites for both the transfected and endogenous miRNAs, Figure 4.1.. Second, as miRNAs may compete with each other for target sites in mRNAs, it may be important to consider RISC saturation in target prediction methodology.

Further, our results have consequences for the development of small RNA therapeutics, considered to hold substantial promise (Stenvang, Lindow et al. 2008). miRNA inhibitors, such as anti-miR-122, have been used to target cholesterol synthesis (Elmen, Lindow et al. 2008; Fabani and Gait 2008) as well as HCV (hepatitis C virus) and HSV2 (herpes simplex virus) (Lupberger, Brino et al. 2008). Therapeutic siRNAs have also been designed for potential treatment of cancer, including in melanoma, against VEGF-A/-C (Shibata, Morimoto et al. 2008), and through anti-miR-21 in glioma (Corsten, Miranda et al. 2007). Our work illustrates the potentially broad consequences of the perturbation of the cellular miRNA activity profile after introduction of si/miRNA inhibitors, and it suggests that these effects be considered quantitatively during development of small RNA therapies. Experiments that quantify the relative concentrations of protein machinery and small RNAs in a particular cellular context, as well as a fuller exploration of the kinetics of the various binding events involved in small RNA biogenesis and function, are clearly required. Our quantitative model implies a procedure for calibrating and potentially avoiding unwanted effects of the designed small RNA therapeutics.

Our work is subject to some limitations. In particular, this report does not attempt to resolve details of the mechanism behind the competition effect. The calculations of the effect, though carefully evaluated in statistical terms, are subject to the inaccuracies of miRNA target prediction, which entails both false positives and false negatives at the level of particular target genes. We therefore argue in terms of overall distributions, rather than attempting to quantify the involvement of individual target sites in transfection-mediated expression changes. In future work, it may be possible to identify quantitative criteria that determine the extent of the competition between exogenous and endogenous miRNAs and their effects on gene targeting. Quantitative detail will depend on knowing the concentration in the cell of the RISC complex and of other components of the small RNA machinery, the concentration of the transfected and endogenous miRNAs, the concentrations of the target mRNAs and the number of actual targets in the cell for a specific small RNA, as well as kinetic parameters such as the on and off rates of small RNAs in the RNA-protein complexes. Models that posit different concentration-dependent and kinetic scenarios could help focus the range of experiments needed to quantify these effects.

Finally, our results may have an important biological correlate, as the competition effect may have a role in normal biological or disease-related cellular processes, such as miRNA-dependent regulatory programs. For example, during both differentiation and disease processes such as cancer, miRNA profiles can change dramatically both in the identity of the dominant miRNAs and in total cellular miRNA concentration. Such changes, via competition for limited resources, may orchestrate observable changes in cellular regulatory programs with potential physiological consequences.

In summary, the observed statistically supported competition effect for small RNAs may point to new biological mechanisms and likely has important practical consequences for the use of small RNAs in functional genomics experiments, development of miRNA target and siRNA off-target prediction methods and development of small RNA therapeutics.

Understanding how RNA perturbation interplays with the cells own gene regulation is critical in order to design optimal therapeutics. Oncogenes are known to be regulated at multiple stages, including post-transcriptionally by microRNAs. We here speculate whether small RNA perturbations could result in a biased effect on oncogenes, which could be the case if oncogenes are more targeted by endogenous miRNAs than other genes. Consistent with this, we find that oncogenes are enriched for microRNAs targets as well as less down-regulated than other target genes after small RNA transfection. We also find that inhibiting microRNAs with antagomirs up-regulates oncogenes less than other genes containing sites for the inhibited microRNA. The observation that the effect on different classes of genes after small RNA transfection depends on levels of endogenous regulation may have implications for the interpretation of experiments as well as for the design of RNA intervention in clinical settings.

5   Chapter Five – Target abundance dilutes microRNA and siRNA activity

5.1   Synopsis

Post-transcriptional regulation by microRNAs and siRNAs depends not only on characteristics of individual binding sites in target mRNA molecules, but also on system-level properties such as overall molecular concentrations. We hypothesize that an intracellular pool of microRNAs/siRNAs faced with a larger number of available target transcripts will down-regulate each *individual* target gene to a lesser extent. To test this hypothesis, we analysed mRNA expression change from 178 microRNA and siRNA transfection experiments in two cell lines. We find that down-regulation of particular genes mediated by microRNAs and siRNAs indeed varies with the total concentration of available target transcripts.

We conclude that to interpret and design experiments involving gene regulation by small RNAs, global properties, such as target mRNA abundance, need to be considered in addition to local determinants. We propose that analysis of microRNA/siRNA targeting would benefit from a more quantitative definition, rather than simple categorization of genes as "target" or "not a target". Our results are important for understanding microRNA regulation and may also have implications for siRNA design and small RNA therapeutics.

All supplementary material referred to in this Chapter can be accessed either from the Nature Molecular Systems Biology website or http://cbio.mskcc.org/ dilution.

5.2   Background

Small RNAs, such as microRNAs and siRNAs, can down-regulate hundreds of target genes. Targeting determinants include many site-specific factors, such as base-pair complementarity (Enright, John et al. 2003; Lewis, Shih et al. 2003; Stark, Brennecke et al. 2003; John, Enright et al. 2004; Lewis, Burge et al. 2005), local context factors (Grimson, Farh et al. 2007; Hammell, Long et al. 2008), and other destabilization signals (Sun, Li et al. 2010). However, systems-level factors also influence small RNA-mediated degradation. For instance, transfected microRNAs and siRNAs cause global de-repression of genes regulated by endogenous microRNAs, most likely through competition for RNA Induced Silencing Complex (RISC) (Khan, Betel et al. 2009). Other systems-level factors include the cellular concentrations of the target transcripts and small RNAs loaded in RISC, determining the kinetics of the regulation. We hypothesize that microRNAs and siRNAs that have a higher number of available targets will down-regulate each *individual* target to a lesser extent than those with a lower number of targets (Fig. 1a, b): We call this the *dilution effect* since those small RNAs with many targets have their effect diluted across many molecules. It follows that the competition between target genes for a limited number of active small RNAs may determine how much a small RNA can down-regulate each of its target mRNAs.

Previous work supports the hypothesis that target abundance can alter small RNA regulation dynamics. Serial dilution experiments in *Drosophila* embryo lysates show that the siRNA-loaded RISC enzyme can be sequestered by competing target molecules (Haley and Zamore 2004). Similarly, sequestration can be artificially induced in living cells by expressing transfected microRNA "sponges" to soak up endogenous microRNA molecules (Ebert, Neilson et al. 2007). Furthermore, siRNAs with a larger number of off-targets show decreased toxicity, perhaps due to a lower effect on each individual gene (Anderson, Birmingham et al. 2008). Finally, there is evidence that target abundance has a significant functional role in an environmental response in plants; namely, phosphate starvation induces a non-coding RNA that sequesters a specific microRNA and de-represses its targets (Franco-Zorrilla, Valli et al. 2007).

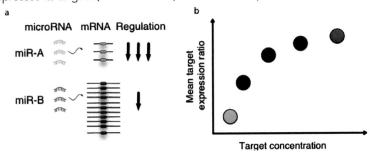

Figure 5.1 Schematic of hypothesis: Mean down-regulation is correlated with target abundance. *(a) Schematic of the hypothesis that target abundance determines mean down-regulation of individual targets. micro/siRNAs with many targets down-regulate their targets to a lesser extent than micro/siRNAs with few targets. (b) Expected correlation between target abundance and log expression ratio. This can also be considered an anti-correlation between down-regulation and target abundance.*

Given this evidence, we reasoned that the effect of microRNAs and siRNAs on each individual gene may be modulated by target transcript abundance. Therefore this effect should be seen in the large number of available small RNA perturbation experiments where genome-wide expression changes are measured.

5.3   Methods

5.3.1   Quantification of transcript abundance

To quantify transcript abundance, we used RNA-seq measurements from HeLa S3 cells (Morin, Bainbridge et al. 2008). We aligned 31bp reads, allowing 2-mismatches, to the reference genome (UCSC genome browser hg18), requiring matches to be unique. This resulted in a total of 17.8 million aligned reads. We examined genes by their AceView 3' UTR annotations(Thierry-Mieg and Thierry-Mieg 2006) and determined the number of reads per nucleotide (RPN), which we use as a proxy for transcript abundance (Mortazavi, Williams et al. 2008). When reads mapped to multiple 3' UTRs, we split the reads evenly between the UTRs. We used alignments to 3' UTRs since most validated microRNA targeting (and siRNA "off-targeting") has been identified in this region.

Alternative measures of transcript abundance included array fluorescence and number of genes targeted (Supplementary Fig 1). Array fluorescence lacks the dynamic range of RNA-seq. That is, probes called as 'present' with low fluorescence frequently seem to be absent from the transcriptome, leaving the problem of determining a proper threshold for minimizing false positives and false negatives (Mortazavi, Williams et al. 2008). The raw number of genes targeted ignores the transcript abundance and thus may sometimes be a poor proxy for target concentration, especially for those small RNAs with few targets. In general, however, these measures are all highly correlated (Supplementary Fig 1). We also used the sum of all individual target sites on all target transcripts, as opposed to just the total number of target transcripts, and came to equivalent conclusions.

To determine the total "predicted target concentration", we sum the abundance for all the predicted targets.

5.3.2   Quantification of target down-regulation

We used microRNA and siRNA transfection experiments followed by microarray assay of gene expression to determine small RNA-mediated down-regulation. We first performed target predictions using heptamer seed matches in the longest 3'-UTR of Refseq annotated genes to determine a set of predicted targets. We also performed target prediction using several other methods: (i) TargetScan conserved targets (Lewis, Burge et al. 2005; Grimson, Farh et al. 2007; Linsley, Schelter et al. 2007); (ii) heptamer seed matches which are conserved in 3 of four distantly related organisms (mouse, rat, chicken, and dog) (Khan, Betel et al. 2009); (iii) heptamer seed match to Aceview 3' UTRs (Thierry-Mieg and Thierry-Mieg 2006); (iv) the miRanda algorithm (John, Enright et al. 2004). All target prediction methods yielded similar results (Supplementary Fig 2).

We examined three measures of down-regulation:
(i)   Average down-regulation (log expression change) of predicted targets.
(ii)   Area between the predicted target cumulative distribution and the background cumulative distribution. Example cumulative distribution functions for targets and background are shown in Figures 1c and 1d. The signed area between these curves yields an estimate for how much down-regulation a mi/siRNA is able to exert.
(iii)   Estimate of the total number of molecules degraded as a percentage of initial number of predicted target molecules. The values for $x(T=1)$ and $x(0)$ are estimated as in the kinetic model and the measure we examine is.
These three methods are highly correlated and yield similar results in our analysis (Supplementary Fig. 1).

5.3.3   Representative set of independent microRNA experiments

There are several datasets that have multiple small RNAs with identical seed sequences. In some cases, the seeds are the same because the entire small RNA is the same. In other cases, the seed is identical but other nucleotides in the small RNA are different. When the entire small RNA is identical, we only include replicate experiments if they were performed by separate labs. When only the seed is identical, we take 2 members of the seed class. When multiple time points are available and we only use one, we take the time point at which the targets are most down-regulated on average (Supplementary Fig. 11).

5.3.4   Down-regulation of primary siRNA targets

We provide evidence suggesting that siRNAs with fewer off-target molecules are more effective at down-regulating their *direct/primary* targets (Fig. 3b). There are several large sets of microarrays where many siRNAs (each separately transfected) are targeted to a single gene. Each of these experiments shows correlation between target abundance and down-regulation; however, only one shows significant correlation (siMAPK14; $P < 0.02$), where as the others show nearly significant correlation. To elucidate the significance of correlation, we combined all the experiments into a single dataset. However, different primary targets are knocked down to different extents on average. To normalize for this effect, we subtracted the mean and divided by standard deviation for each set of siRNAs targeting a single gene. That is, for each panel of "raw data" in Supplementary Fig. 9, we subtract the mean of the y-axis and divide by the standard deviation of the y-axis. We do not change the x-axis. The resulting transformed data is shown in Fig. 3b and Supplementary Fig. 9. The transformed data shows highly significant correlation.

Note that the siRNAs targeted to MAPK14 and SOD1 have a single nucleotide mismatch to the target in HeLa cells. While MAPK14 has a perfect match siRNA, we do not use it since all other siRNAs have a single nucleotide mismatch. Inclusion of the perfect match siRNA does not alter conclusions.

5.4   Results and Discussion

5.4.1   Target abundance affects average down-regulation by small RNAs

The mean down-regulation of target mRNAs for different transfected small RNAs is highly variable. For instance, when miR-155 and miR-128 are separately transfected into HeLa cells, the targets of miR-155 are more down-regulated than the targets of miR-128 (Fig. 1c,d). We hypothesized that this variability is due, in part, to variable abundance of target mRNAs. To examine this phenomenon more systematically, we analyzed the mRNA expression changes after transfection for a panel of 146 small RNA transfection experiments in HeLa cells and 32 small RNA transfection experiments in HCT116 cells (Lim, Lau et al. 2005; Jackson, Burchard et al. 2006; Schwarz, Ding et al. 2006; Grimson, Farh et al. 2007; He, He et al. 2007; Kittler, Surendranath et al. 2007; Linsley, Schelter et al. ; Anderson, Birmingham et al. 2008; Selbach, Schwanhäusser et al.) (Supplementary Table 1).

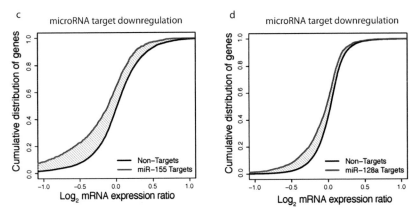

Figure 5.2 **Differential down-regulation of genes after (a) miR-155 and (b) miR-128 transfections**

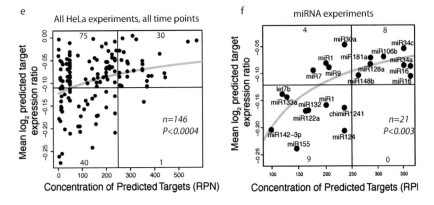

Figure 5.3 **Mean down-regulation is correlated with target abundance** *(a) Predicted target concentration and mean log expression ratio across 146 micro/siRNA transfection experiments. (b) Predicted target concentration and mean down-regulation across 21 independently measured, single time point microRNA transfection experiments. Curves were fit to log(1-a/(x-b)), where a and b were determined by least square error.*

To analyze the proposed dilution effect, one has to quantify the abundance and mean down-regulation of target transcripts for each small RNA transfection. Target mRNA abundance in HeLa cells is quantified using RNA-seq data in units of reads per nucleotide (RPN) (Morin, Bainbridge et al. 2008; Mortazavi, Williams et al.). Down-regulation is computed from the ratio of expression levels after transfection to the levels before transfection, where the expression levels are measured by hybridization on microarrays.

We examined all 146 HeLa experiments and found a significant correlation (P < 0.0004, Pearson's r=0.31, Spearman's ρ=0.30) between the concentration of predicted targets of the transfected small RNA and the average log expression ratio of the target mRNAs (Figure 5.3).

This can also be considered an anti-correlation between abundance of predicted targets and mean down-regulation. This anti-correlation reflects the increased dilution of a limited pool of transfected small RNAs over an increased number of available target sites on targeted transcripts.

To determine the match between small RNAs and their targets, we used 7mer seed matches in the 3' UTR of mRNAs. We explored several different ways of quantifying abundance and down-regulation and of predicting targets, with similar results (Methods, Figures 5.4 and 5.5, Supplementary Tables 2 and 3).

An alternative statistical measure of the relation between low target abundance and high down-regulation can be obtained by dividing the experiments into quadrants by mean log expression ratio and target concentration (Figure 5.3a; cutoff for mean log expression ratio is -0.11 and cutoff for predicted target concentration is 250RPN). Strikingly, when target abundance is *low*, the fraction of experiments with *high* down-regulation is 10 times larger than when target abundance is *high* (40/115 = 34.8%, 1/31 = 3.2%; P < 0.0003, Fisher's exact).

Figure 5.4 Alternative methods for quantifying down-regulation and target ▶ abundance. *(a) Mean down-regulation and area between cumulative distribution curves are highly correlated. (b) Mean down-regulation and percentage of target molecules degraded are highly correlated. (c) Predicted target abundance by RNA-seq (RPN) and array fluorescence are highly correlated. The arrays used by Anderson et al. have different range and are shown separately from other experiments. (d) Predicted target abundance by RNA-seq (RPN) and number of genes targeted are highly correlated.*

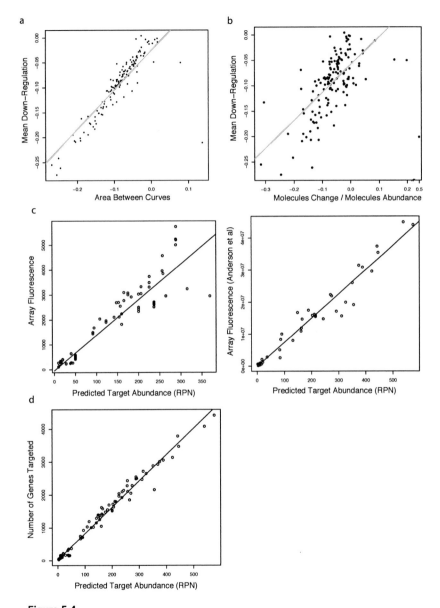

Figure 5.4

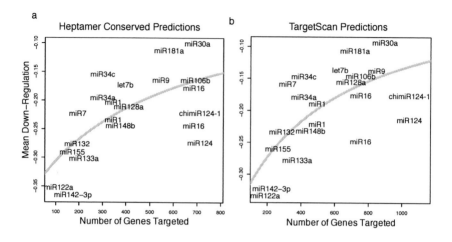

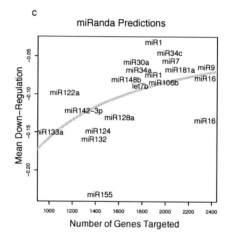

Figure 5.5 Different target predictions methods yield similar results. *The methods examined include (a) conserved heptamers, (b) TargetScan, and (c) miRanda*

5.4.2 Down-regulation by transfected microRNAs is a function of target abundance

Up to this point, the analysis has examined both siRNA and microRNA transfection experiments pooled together. We then wanted to examine, whether the effect was in any way different for the subset of transfection experiments using microRNAs. We found that a representative set of microRNA experiments (Methods) show a significant rank correlation between target abundance and mean log expression ratio (Fig. 1f, $P < 0.003$, $\rho=0.62$, n=21). Beyond correlation measures, we also quantify the dynamic range of the dilution effect. For instance, miR-142-3p, the microRNA with the lowest target abundance can down-regulate its targets 83% more (on average) than miR-16, the microRNA with the highest target abundance.

We analyzed 32 transfection experiments in a second cell type, HCT116 (dicer-/- or dicer+/+) and found that high mean down-regulation is associated with low predicted target abundance ($P < 0.054$, Fisher's exact, Figure 5.6, Supplementary Material). Furthermore, the fraction of transfections that induce large down-regulation is very similar in HCT-116 as it was in HeLa. Namely, 33% (n=4) of the transfection experiments whose small RNA had low target-abundance (n=12) resulted in predicted targets being highly down-regulated, whereas there was only one exception experiment whose small RNA had high down-regulation when there was high abundance of predicted targets (n=20).

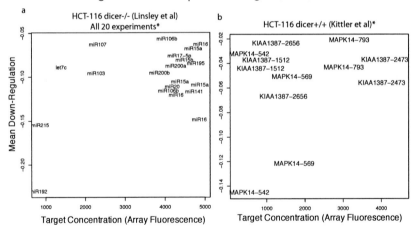

Figure 5.6 Transfection experiments in HCT-116 cells show mean log expression ratio is correlated with concentration of predicted targets, as estimated using array fluorescence.

Since microRNAs can inhibit translation whilst showing little or no change in mRNA expression (Filipowicz, Bhattacharyya et al. 2008), we tested whether target mRNA abundance had an effect on the amount of *protein* down-regulation.

Using mass spectrometry measurements after microRNA transfection into HeLa cells (Selbach, Schwanhäusser et al. 2008), we found mean down-regulation of protein products is consistent with our hypothesis (ρ =0.60, n=5, Supplementary Fig 4). Though the correlation does not reach statistical significance due to low sample size, the trend is similar to the significant dilution effect in mRNA micro-array experiments. This trend is relevant as there is no guarantee that changes in mRNA are reflected in the changes in functional protein.

Our results establish a significant anti-correlation between mean down-regulation and target abundance; however, this could be the result of other variables that correlate with down-regulation as well as target abundance. We explored three possible predictors of down-regulation: A+U content, 3'-UTR length, and expression level of individual transcripts. Neither the A+U content of the seed nor of the entire microRNA are significantly correlated with down-regulation (at a threshold of 0.05; Supplementary Fig. 5). The 3' UTR length is weakly correlated with target abundance and mean log expression ratio (Supplementary Fig. 6 and 7) and individual genes with higher expression levels tend to be more down-regulated on average when microRNAs are transfected (Supplementary Fig. 8). We have shown that the A+U content is not correlated with mean down-regulation, but the other two predictors might contribute to the dilution effect. However, further analysis can control for these predictors, as we describe next.

5.4.3   Pairwise analysis controls for artifacts

To test our hypothesis while controlling for the potential biases, increase statistical power, and examine differential regulation for *individual genes*, we performed a second type of analysis. We determined the expression changes of all genes that are targeted by any pair of microRNAs and compared the difference in average down-regulation with respect to the difference in total number of targets transcripts (Figure 5.7).

This shared target comparison controls for various sources of bias, including A+U content and 3' UTR length distributions. Figure 5.8 shows a table of selected genes where each gene is targeted by both miR-A, which has fewer total targets in the cell, and miR-B, which has more. For example, *Smad5* and *Tgfbr2* are far less down-regulated when miR-106 is transfected compared to miR-155; similarly *Nfat5* is down-regulated much less with miR-16 when compared to miR-122 transfections (Figure 5.8). We also found a highly significant correlation between the difference in target abundance and difference in down-regulation (Fig. 2c, $P < 10^{-15}$, $\rho=0.59$). We determine an empirical p-value of $P < 10^{-5}$, which uses an empirical background distribution using a set of randomly selected non-targets the same size as the set of shared targets (Figure 5.9). Finally, to control for possible lab-specific artifacts, we specifically examined microRNA pairs from a single lab and found similar results (nominal $P < 10^{-6}$, $\rho=0.74$, Supplementary Material).

A

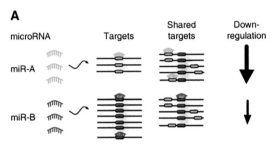

Figure 5.7 Schematic of pairwise analysis: *Given miR-A and miR-B, which both target the same gene, they will regulate differentially in proportion to their individual overall target abundance*

B

Genes	miR-A	miR-A Reg	miR-B	miR-B Reg
Smad5	miR-155	−1.29	miR-106	−0.1
Tgfbr2	miR-155†	−1.27	miR-106††	−0.07
Nfat5	miR-122†	−3.65	miR-16	−0.74
Tgfbr1	miR-133†	−0.64	miR-181†	−0.05
Foxq1	miR-133†	−1.75	miR-128†	−0.81
Hmga1	let-7†	−1.47	miR-16††	−0.85
Smad3	miR-155	−0.98	miR-16†	−0.19

† represents a conserved site
†† represents two conserved sites

Figure 5.8 Examples of genes that may be differentially regulated (log$_2$ expression ratio) as a result of total target abundance

The full set of microRNA pairs considered in our calculations together with differential down-regulation of shared targets are provided in Supplementary Table 4. In summary, this more refined analysis controlled for potentially confounding variables and confirmed the dilution effect for microRNAs.

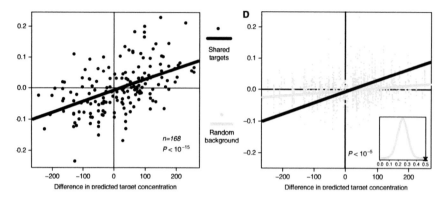

Figure 5.9 Analysis of shared targets of microRNAs. *(a) All pairs of microRNAs are used to search for shared targets. Each point is the difference in mean down-regulation and difference in target abundance for a given pair of microRNAs. (b) Background distribution (grey) as a result of randomizing targets. Inset shows empirical p-value of $P < 10^{-5}$. Real correlation line is black.*

5.4.4   Target abundance affects primary target and off-target average down-regulation by siRNAs

We also show that siRNA off-target down-regulation is correlated with on- and off-target abundance. First, we found highly significant correlation between mean log expression ratio of off-targets and off-target abundance (Figure 5.X, $P < 0.0001$, $\rho=0.37$). However, siRNAs are designed to degrade a single *primary* target gene. Therefore, we analyzed correlation between each siRNA's down-regulation of its primary target and abundance of off-targets.

We normalized the down-regulation by each siRNA with the same primary target by subtracting the mean and dividing by standard deviation, since different primary targets can be knocked down with highly different efficiencies (Supplementary Fig. 9, Supplementary Material). We found a significant rank correlation between log expression ratio of primary target and abundance of off-targets (Fig. 3b, $P < 0.007$, $\rho =0.34$). This indicates that off-target abundance should be considered in the design of siRNA to maintain high knockdown efficiency.

Recent work has noted that siRNAs with many off-targets may reduce RNAi-induced toxicity (Anderson et al, 2008). However, if this strategy were employed, our results suggest one would face a trade off between reduced siRNA toxicity (more off-targets may lead to decreased toxicity) and increased knockdown of direct siRNA target (fewer off-targets lead to increased knockdown).

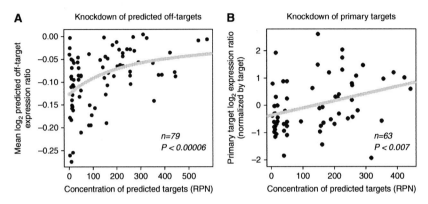

5.10 Target abundance determines siRNA primary target and predicted off-target down-regulation. *(a) Off-targets for a siRNA experiments show significant correlation. Curve was fit to log(1-a/(x-b)), where a and b were determined by least square error. (b) The log expression ratio of the primary siRNA target is correlated with predicted off-target concentration. The down-regulation for each primary target is normalized such that multiple siRNAs for different targets can be compared.*

5.4.5 Michaelis-Menten kinetics describes total transcripts degraded

Since we found a significant anti-correlation between down-regulation and total target abundance, we wanted to explore the kinetics that determines the total number of mRNA transcripts degraded. Given the pre-transfection target transcript concentration $x(0)$, we computed the post-transfection concentration $x(T)$, at T=1 day, as follows. We used RNA-seq target transcript abundance for $x(0)$ and determined post-transfection abundance using log expression ratio after transfection from microarray experiments: if we denote $x_i(0)$ as the pre-transfection abundance of target gene i and $\frac{y_i(T=1)}{y_i(0)}$ as the change in expression, then we obtained an estimate for the post-transfection abundance for each target i as

$$x_i(T=1) \approx \frac{y_i(1)}{y_i(0)} x_i(0) \quad (1)$$

We then estimated the *initial* velocity, i.e. time rate of decrease of transcript concentration, as $v = x(0)-x(T=1)$, in order to express the velocity as a function of the initial target concentration using Michaelis-Menten kinetics. Our assumption is that T=1 day is sufficiently early to approximate the initial velocity.

We computed v for each of the 146 transfection experiments in HeLa. Empirically, v is significantly dependent on target concentration and fits the Michaelis-Menten model better than linear or constant models (Supplementary Fig. 10; Supplementary Material). That is, we can fit values of V_{max} and K_m such that $v =V_{max}x(0) / (K_m + x(0))$ and express:

$$\log\left(\frac{x(T=1)}{x(0)}\right) = \log\left(\frac{x(0) - V_{max}x(0)/(x(0) + K_m)}{x(0)}\right) = \log\left(1 - V_{max}/(x(0) + K_m)\right). \quad (2)$$

Therefore, this kinetics can be used to predict the expected log expression change of target genes as a function of target abundance.

This relationship provides an explanation of the observed anti-correlation between mean down-regulation and concentration of predicted targets. These results motivate a more quantitative framework for studying small RNA regulation and kinetics.

5.4.6   Conclusions

We find evidence that the activity of a small RNA can depend on the abundance of its targets and potential for crosstalk between targets via dilution of microRNAs/siRNAs. This may have implications for research and clinical applications of siRNA technology and help explain endogenous microRNA dynamics. .

In the clinic, our observed dilution effect may have important implications for the design and efficacy of therapeutic siRNAs, suggesting a key design criterion for therapeutic siRNAs is minimizing the number of off-target sites on highly expressed genes. This may improve efficient direct target down-regulation at moderate RNAi concentrations and thereby avoid unwanted saturation effects (Khan et al., 2009). In research applications, siRNAs, such as those used in functional genomic screens, will be more likely to function optimally in cells where off-transcript numbers are low. This may be particularly relevant when designing constructs for cell types that express very few genes at very high levels, such as hepatocytes (Ramsköld, Wang et al. 2009).

Endogenous microRNAs may have their effects diluted by highly abundant target transcripts in particular cell types or states. The activity of endogenous microRNAs on specific targets may be significantly altered in contexts where target concentrations change dramatically, such as shifts in environment during differentiation, development, and evolution. Specifically, particular pairs of microRNAs and target genes may be functionally constrained to co-evolve to maintain a constant strength of down-regulation. This is different from (though not inconsistent with) the anti-target hypothesis (Farh, Grimson et al. 2005; Stark, Brennecke et al. 2005) in which mRNAs that dominate a cell type are thought to specifically avoid microRNA targeting.

Finally, quantifying down-regulation suggests a departure from considering microRNA targets based on RNA sequence alone. Imbalances in the relative concentrations of microRNAs and gene targets might exaggerate or compensate for sequence mismatches between the two species. For example, highly expressed microRNAs with *very few* target transcripts may well effectively down-regulate mRNAs with 'weak sites', such as those containing G:U wobbles. Conversely, weakly expressed microRNAs with *many* potential mRNA target transcripts, may fail to down-regulate species with excellent sequence matches. This will be important to consider in a new generation of microRNA target prediction methods.

Our work demonstrates that small RNA activity can be influenced by the concentrations of the target mRNA molecules, a phenomenon that may alter the effectiveness of therapeutic, investigational and endogenous small RNAs.

6   Chapter Six – Conclusions and Future work

6.1   Synopsis

First I will summarize the main results of the thesis, followed by an outline of 4 future research projects, which address microRNAs as part of the cellular system of gene regulation and gene dose homeostasis. Developments in understanding RNA:RNA interactions and the effects on cellular phenotypes has still a long way to go. The last part of this thesis attempts address some of the outstanding issues in this area, which I believe need a different kind of approach to the one taken over the past 7 years.

6.2   Summary of Thesis

This thesis can be divided into three sections:

6.2.1.1   The discovery of microRNAs, microRNA target prediction in Drosophila melanogaster

Covering Chapter 2. The first part describes the work I did developing the bioinformatic component of the discovery of new microRNAs in *Drosophila* melanogaster. The pipeline went on to be used in discovery of microRNAs in *Danio rerio* (ref). Chapter 2 describes the first systematic attempt to identify microRNAs in *Drosophila* over different development stages. We identified over 70 candidate microRNAs, most with no known function.

Chapter 2 describes the first systematic attempt to identify microRNAs in *Drosophila*. We identified over 70 candidate microRNAs, most with no known function. We then asked whether computational methods could be used to predict likely gene targets and hence provide clues to function. I describe the miRanda algorithm for identifying microRNA targets developed to identify target genes of *Drosophila* microRNAs. The precise rules and energetics for pairing between a miRNA and its mRNA target sites, with probable involvement of a protein complex, were not known and could not be deduced from the few experimentally proven examples. Therefore, any computational methods for the identification of potential miRNA target sites were at risk of having a substantial rate of false positives and false negatives. Based on analysis of the known examples, we biased our method toward stronger matches at the 5' end of the miRNA, and used energy calculation plus conservation of target site sequence to provide our current best estimate of biologically functional matches. The program and algorithm were optimized to be flexible for the user. First, the user could define the threshold cutoff for accepting a site, second the conservation consideration can uncoupled so users can decide whether or not they requite conservation, and third, the program was stand alone so that the user could run on any sequence of their choice. I then present the results as applied to mRNAs in *Drosophila melanogaster*. Our results suggest that miR-NAs target the control of gene activity at multiple levels, specifically transcription, translation and protein degradation, in other words, that miRNAs act as meta-regulators of expression control. Among biological processes, we find that the most prominent targets include signal transduction and transcription control in cell fate and developmental timing decisions, as well as morphogenetic processes such as axon guidance. These processes share the need for the precise definition of boundaries of gene activity in space and time. Our findings therefore support and expand earlier work on the role of miRNAs in developmental processes [42, 80]. In addition, we predict that miRNAs also play an important role in controlling gene activity in the mature nervous system.

6.2.1.2   *miRanda, microRNA target and function prediction in human*

The second main part of the thesis (Chapters 3–7) describes work which addresses the potential scope of regulation by microRNAs in animals. Since we only knew individual targets in 2003, a total of five, the size of the breadth of microRNA regulation was unknown. The question was; can we use current knowledge on the four targets to predict microRNA function. Did microRNAs have one, a few key functional targets or a program dependent on affecting hundreds of genes? This could not be addressed experimentally at the time as no high throughput technique existed which could identify multiple targets. One way to address this is to try to predict what genes might be target through base pairing to the mRNA, and this is what we did. Of course, this approach makes the general assumption that genes which pairing causes a change in protein dose, has an effect on phenotype.

A goal of our computational approach was to put bounds on the number of genes potentially regulated by a specific microRNA or family of microRNAs. Our approach was based in the knowledge from the few known microRNA-target relationships, that microRNAs target mRNAs through imperfect base-pairing with the 3'UTR. It is the nature of the 'imperfectness' that provides the computational challenge to predict the matching functional microRNA –mRNA pairs. When we started this work only five microRNA targets were known and only one of these in *Drosophila* (bantam and hid). The characteristics of the base pairing, such as internal bulges and mismatches and wobbles, made it important to design a user-flexible approach to the problem. The ability to add new rules, which were inevitable come from individual experiments, made it necessary that the method also be modular.

Chapter three extends the algorithm miRanda to predict the scope of regulation of genes by microRNAs in **humans** and other sequenced vertebrates. We develop the miRanda algorithm by incorporating rules distilled from new target validation experiments. We showed that over 4,500 genes have one or more miRNA target sites in their 3' UTRs conserved in mammals at 90% target site conservation. This means we predicted a minimum of 20% of human protein repertoire is regulated by microRNAs. We found a large range of specificity for miRNAs and suggesting that regulation of one message by one miRNA is rare. The full set of conserved target genes between fish and mammals indicates a wide functional range of conserved targets. This Chapter concluded that a lower bound of the number of actively regulated genes was 10-20% of the total number and hence that about 1% of genes (miRNAs) control the expression of more than 10% of genes. The end of Chapter Three I extrapolate to predict over 60 % of human protein coding genes are under some endogenous microRNA regulation. The question remains by how mow much microRNA targeting contributes to the mRNA half-life of each target gene and this will be discussed below in future directions.

As miRNA and mRNA have to be present simultaneously at minimum levels in the same cellular compartment for a biologically meaningful interaction, more precise expression data as a function, for example, of developmental stage [42], will be extremely useful and will be incorporated in future versions of target prediction methods. Similarly, further work will include the analysis of potential target sites in coding regions and 5' UTRs, as well as conservation and adaptation of target sites in many species.

This genome-wide scan for potential miRNA target genes gives us a first glimpse of the complexity of the emerging network of regulatory interactions involving small RNAs (see Additional data). Both multiplicity (one miRNA targets several genes) and cooperativity (one gene is targeted by several distinct miR-NAs) appear to be general features for many miRNAs, as already apparent with the discovery of the targets for *lin-4* and *let-7*. The analogy of these many-to-one and one-to-many relationships to those of transcription factors and promoter regions is tempting and elucidation of the network of regulation by miRNAs will make a major contribution to cellular systems biology. In the meantime, we would not be surprised if experiments focusing on target candidates filtered in this way have a high rate of success and help to unravel the biology of regulation by miRNA-mRNA interaction

6.2.1.3 A system-level approach to microRNA function

Chapters Four and Five, take a more system level approach to the problem of predicting and understanding microRNA functions and siRNA targeting. The thesis is that global cellular properties such as concentrations of components will affect the targeting capacity of these small RNAs. In Chapter 7 we show that *endogenous* microRNA regulation is attenuated after *exogenous* small RNA over-expression or inhibition. This results in hundreds of genes being unexpectedly upregulated, Chapter 4.

In Chapter 5 we show how mRNA target abundance attenuates microRNA and siRNA activity. We hypothesized that an intracellular pool of microRNAs/siRNAs faced with a larger number of available target transcripts will down-regulate each *individual* target gene to a lesser extent. To test this hypothesis, we re-analysed mRNA expression change from 178 microRNA and siRNA transfection experiments in two cell lines. We find that down-regulation of particular genes mediated by microRNAs and siRNAs indeed varies with the total concentration of available target transcripts. We conclude that to interpret and design experiments involving gene regulation by small RNAs, global properties, such as target mRNA abundance, need to be considered in addition to local determinants. We propose that analysis of microRNA/siRNA targeting would benefit from a more quantitative definition, rather than simple categorization of genes as "target" or "not a target". Our results are important for understanding microRNA regulation and may also have implications for siRNA design and small RNA therapeutics.

6.3   Future directions

6.3.1   Quantitative, functional target prediction

With "state of the art" 2009 target prediction methods, eight times more genes are predicted to be down-regulated than actually are (false positives), Figure 6.1.

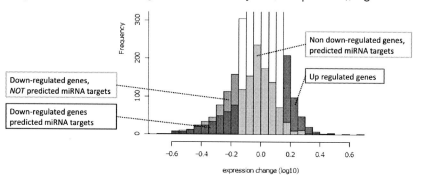

Figure 6.1

Future work should concentrate on improving this significantly as opposed to incrementally. We have shown that system-level cellular properties can affect the level of microRNA and siRNA activity. The next step is to incorporate these analyses into predicting the amount of down-regulation of a microRNA has on a particular transcript. For instance target abundance could be a feature in a machine learning approach to target prediction.

New experimental techniques will change the way we analyze small RNA targeting. Recent experimental work attempts to find microRNA binding sites more directly. Since microRNAs are bound to members of the AGO protein family when they regulate mRNAs, methods have been developed which use UV cross-linking (Greenberg 1979) the labeled AGO-microRNA with the bound mRNA. This is followed by the complex immunoprecipitation (dreyfuss 1984 Maynard 1981), followed by the isolation of crosslinked RNA segments and either microarray detection (Brown Hendrickson, Hannon) or cDNA sequencing (CLIP) (Ule et al., 2003). An improved method for isolation of the mRNA short sequences bound by AGOs (and other RNA-BPs) is referred to as PAR-CLIP (Photoactivatable-Ribonucleoside-Enhanced Crosslinking and Immunoprecipitation) (Tuschl Hafner 2010). However, although the AGO bound mRNAs are 'enriched' for the seed sequences, over X% of unbound sequences contain endogenous seed sequences (using equivalent set of microRNAs). This suggests that the problem of target recognition is somewhat unsolved.

My hypothesis is that we will have to consider global cellular properties to be successful, such as, mRNA abundance, mRNA stability and combinatoric regulation, synergistic or otherwise with the thousands of RNA binding genes in a cell.

6.3.2   Combinatorics of RNA binding proteins and microRNAs in gene regulation

MicroRNA/siRNA targeting may be modulated by other mRNA sequence elements such as binding sites for the hundreds of RNA binding proteins expressed in any cell. Messenger RNAs are bound by multiple members of the hundreds of RNA binding proteins (Finn et al. 2008) in the human genome. Though potentially reminiscent of transcriptional regulation, the regulation of mRNA stability is a nascent field. Many of these RNA binding proteins regulate mRNAs via sequences particularly in the 3'UTRs. ARE-mediated and microRNA mediated regulation have been associated in at least four studies. Cases of cooperative (Jing et al. 2005; Kim et al. 2009) and competitive (Bhattacharyya et al. 2006) binding of an ARE-BP have been described, and miRNA-loaded RISC in 3'UTRs and AU-rich motifs have been found to enhance miR-223 binding in the RhoB 3'UTR (Sun et al. 2009). Intriguingly, ARE-BPs, miRNAs and components of RISC all co-localize in p-bodies, cellular sites of mRNA decay and translational repression (Franks and Lykke-Andersen 2007; Eulalio et al. 2008). Further highlighting the potential cross-talk between the miRNA and ARE pathway is the observation that the RISC complex includes two ARE-associated proteins, PAIP1 and FXR1 (Caudy et al. 2002; Peng Jin et al. 2004), and that the ARE binding protein KSRP, besides regulating mRNA stability, is a component of both Drosha and Dicer complexes regulating biogenesis of a subset of miRNAs (Trabucchi et al. 2009). Finally the Steitz lab has shown that the ARE-mediated stability control can also function together with microRNAs in RISC to stabilize mRNAs in certain cellular conditions (Vasudevan et al. 2007).

Re-analysis of the hundreds of expression experiments after small RNA perturbation will identify potential cis-regulatory motifs which correlate or anti-correlate with the small RNA targeted down-regulation. Motifs recognized by HuR and other AU-rich element (ARE) binding proteins have already been associated with microRNA regulation (refs). However, to date no-one has done a systematic computational screen for motifs enriched in mRNAs whose expression changes after RNAi, which may reveal both de novo and known motifs, already associated with one or more RNA binding protein. Figure 9.1 is a schematic representing synergy of motif with microRNA or siRNA binding sites

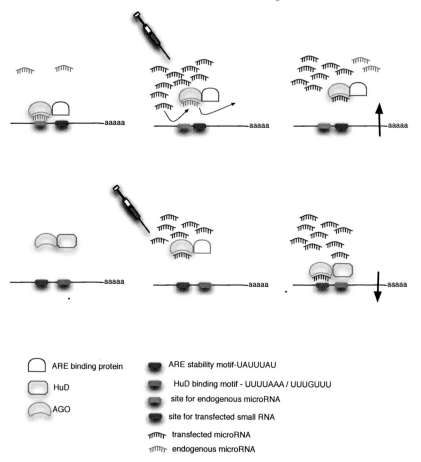

⬭ ARE binding protein	⬛ ARE stability motif-UAUUUAU
⬯ HuD	⬛ HuD binding motif - UUUUAAA / UUUGUUU
⬮ AGO	⬛ site for endogenous microRNA
	⬛ site for transfected small RNA
	ᴨᴨᴨ transfected microRNA
	ᴨᴨᴨ endogenous microRNA

Figure 6.2

These include the CPE (Piqué et al. 2008), FMR (J C Darnell et al. 2001; Schaeffer et al. 2001) and AU-rich elements (ARE) (reviewed in (Barreau et al. 2005)). AREs recruit binding proteins (ARE-BPs) that signal rapid degradation or increased stability of mRNAs in response to stress or developmental cues (Barreau et al.

2005). For instance, the mRNAs of TNFalpha contain a conserved ARE that ex-
erts tight post-transcriptional control (Vasudevan and Steitz 2007). A core motif
in many AREs is the heptanucleotide UAUUUAU and studies have shown that

6.3.3   Do microRNAs as a whole regulate specific kinds of genes more than others?

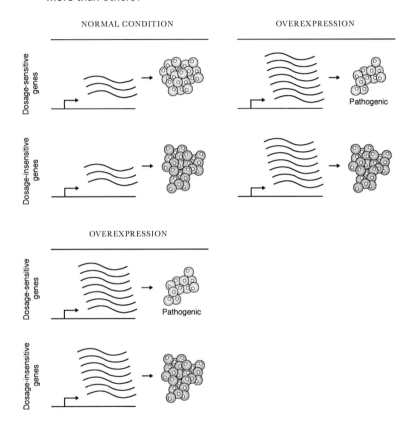

Surprisingly, the expression level of most genes may fluctuate physiologically
across cells without causing any harmful effects. On the other hand,
groups of genes are known to be particularly sensitive to overexpression and
dysregulation may be pathogenic. Examples of such "dosage-sensitive"
genes are oncogenic kinases, which can contribute to tumorigenesis if overex-
pressed.

MicroRNAs can tune the expression levels of genes, but paradoxically such regulation levels can be relatively small compared to the natural fluctuation of gene expression. It is known that dosage-sensitive genes are tightly regulated at multiple stages. If this is the case, we may expect to see dosage-sensitive genes enriched for microRNA regulation and that this phenomena should be evolutionarily conserved. To test this hypothesis one could investigate the relationship between the dosage-sensitivity and the number of microRNA target sites in a gene. One would expect this to be evolutionarily conserved, so the hypothesis could be tested across many different animals.

Cancer-associated genes are known to be regulated at multiple stages, including post-transcriptionally by microRNAs. We here speculate whether small RNA perturbations could result in a biased effect on cancer-associated genes, which could be the case if these are more targeted by endogenous miRNAs than other genes.

6.3.4   Therapeutics and siRNAs – the future

RNA interference (RNAi) has huge potential for therapeutic applications. In a remarkably short time since its discovery, RNAi has already entered human clinical trials in a number of disease areas. However, rapid acceptance of the use of RNAi has been accompanied by recognition of a number of hurdles for the technology, including a lack of specificity. Off-target activity can complicate interpretation of phenotypic effects following gene-silencing experiments and can potentially lead to unwanted toxicities following therapeutic administration of siRNAs. Early reports in the literature describe siRNAs as being exquisitely specific 1-6, demonstrating that a target gene can be silenced by a complementary siRNA but not an unrelated siRNA 2, 6, that synthetic siRNAs silence their target but do not induce an immune response 2, and that a single mismatch at the cleavage site of the siRNA abolished target silencing 1, 3, 4. However, each of these studies investigated one or a few target genes, which has lead to questions about whether unbiased, global expression profiling would reveal previously unsuspected off-target activity. Although initial studies with microarray expression profiling appeared to support the idea that siRNAs were specific for silencing of the intended target 7, 8, it is now recognized that a number of properties of siRNAs and/or the technologies used to deliver them, lead to sub-optimal specificity (Figure 1).

The off-target effects fall into three broad categories. First, the microRNA like effect of the sequnce ie partial matching to targets The siRNA

(ii) siRNAs and/or the vehicles used to deliver them can trigger an inflammatory response through activation of toll-like receptors designed to recognize nucleic acid and cellular material from invading pathogens, and mount a defensive response and (iii) Exogenous siRNAs can saturate the endogenous RNAi machinery, producing widespread effects on microRNA processing and function.

6.4 Summary

This thesis describes work from the time when mciroRNAs were discovered, and a first microRNA target prediction mehtod was developed, miRanda. This is followed by a more system approach which discovers that global properties of the cell affect individual microRNA-gene relationships. This approach explains some puzzzles about the cellualr responses to exogenous microRNAs and siRNAs and how microRNA.siRNA regulation of genes effectively causes cross talk between mRNAs. Both the saturation and competition reulsts in Chapters 4 and 5 have important consequences for basic quantitative understanding of small RNA taregting, the development of effective siRNAs for fucnitonal genomics. Importantly these coomptetitive effects also have consequences for small RNA therapeutics, in particular the work specifies which genes may be upregulated unexpectedly when delivering an microRNA/siRNA mimic or using a microRNA antagonist to inhibit an endogenous microRNA. The final chapter outlines some future projetcs which naturally follow the work presented in the thesis.

THE END

7   References

(2002). "The FlyBase database of the Drosophila genome projects and commu-nity literature." Nucleic Acids Res **30**(1): 106-8.

Abrahante, J. E., A. L. Daul, et al. (2003). "The Caenorhabditis elegans hunch-back-like gene lin-57/hbl-1 controls developmental time and is regulated by microRNAs." Dev Cell **4**(5): 625-37.

Akam, M. (1998). "Hox genes: from master genes to micromanagers." Curr Biol **8**(19): R676-8.

Amara, F. M., A. Junaid, et al. (1999). "TGF-beta(1), regulation of alzheimer amy-loid precursor protein mRNA expression in a normal human astrocyte cell line: mRNA stabilization." Brain Res Mol Brain Res **71**(1): 42-9.

Ambros, V. (2004). "The functions of animal microRNAs." Nature **431**(7006): 350-5.

Ambros, V., B. Bartel, et al. (2003). "A uniform system for microRNA annota-tion." Rna **9**(3): 277-9.

Ambros, V., R. C. Lee, et al. (2003). "MicroRNAs and other tiny endogenous RNAs in C. elegans." Curr Biol **13**(10): 807-18.

Anderson, E., A. Birmingham, et al. (2008). "Experimental validation of the im-portance of seed complement frequency to siRNA specificity." RNA **14**(5): 853.

Antar, L. N. and G. J. Bassell (2003). "Sunrise at the synapse: the FMRP mRNP shaping the synaptic interface." Neuron **37**(4): 555-8.

Aravin, A. A., G. J. Hannon, et al. (2007). "The Piwi-piRNA pathway provides an adaptive defense in the transposon arms race." Science **318**(5851): 761-4.

Aravin, A. A., M. Lagos-Quintana, et al. (2003). "The small RNA profile during Drosophila melanogaster development." Dev Cell **5**(2): 337-50.

Aravin, A. A., N. M. Naumova, et al. (2001). "Double-stranded RNA-mediated silencing of genomic tandem repeats and transposable elements in the D. melanogaster germline." Curr Biol **11**(13): 1017-27.

Ashburner, M. and S. Lewis (2002). "On ontologies for biologists: the Gene On-tology--untangling the web." Novartis Found Symp **247**: 66-80; discussion 80-3, 84-90, 244-52.

Baek, D., J. Villen, et al. (2008). "The impact of microRNAs on protein output." Nature.

Baker, K. D., J. T. Warren, et al. (2000). "Transcriptional activation of the Drosophila ecdysone receptor by insect and plant ecdysteroids." Insect Biochem Mol Biol **30**(11): 1037-43.

Bartel, D. P. (2004). "MicroRNAs: genomics, biogenesis, mechanism, and function." Cell **116**(2): 281-97.

Bartel, D. P. (2009). "MicroRNAs: target recognition and regulatory functions." Cell **136**(2): 215-33.

Bashirullah, A., A. E. Pasquinelli, et al. (2003). "Coordinate regulation of small temporal RNAs at the onset of Drosophila metamorphosis." Dev Biol **259**(1): 1-8.

Bender, L. B., P. J. Kooh, et al. (1993). "Complex function and expression of Delta during Drosophila oogenesis." Genetics **133**(4): 967-78.

Beneyto, M. and J. H. Meador-Woodruff (2008). "Lamina-specific abnormalities of NMDA receptor-associated postsynaptic protein transcripts in the prefrontal cortex in schizophrenia and bipolar disorder." Neuropsychopharmacology **33**(9): 2175-86.

Berezikov, E., E. Cuppen, et al. (2006). "Approaches to microRNA discovery." Nat Genet **38 Suppl**: S2-7.

Berezikov, E., F. Thuemmler, et al. (2006). "Diversity of microRNAs in human and chimpanzee brain." Nat Genet **38**(12): 1375-7.

Betel, D., M. Wilson, et al. (2008). "The microRNA.org resource: targets and expression." Nucleic Acids Res **36**(Database issue): D149-53.

Bray, N., I. Dubchak, et al. (2003). "AVID: A global alignment program." Genome Res **13**(1): 97-102.

Brennecke, J., D. R. Hipfner, et al. (2003). "bantam encodes a developmentally regulated microRNA that controls cell proliferation and regulates the proapoptotic gene hid in Drosophila." Cell **113**(1): 25-36.

Brennecke, J., C. D. Malone, et al. (2008). "An epigenetic role for maternally inherited piRNAs in transposon silencing." Science **322**(5906): 1387-92.

Brennecke, J., A. Stark, et al. (2005). "Principles of microRNA-target recognition." PLoS Biol **3**(3): e85.

Britten, R. J. and E. H. Davidson (1969). "Gene regulation for higher cells: a theory." Science **165**(891): 349-57.

Britten, R. J. and E. H. Davidson (1971). "Repetitive and non-repetitive DNA sequences and a speculation on the origins of evolutionary novelty." Q Rev Biol **46**(2): 111-38.

Broadus, J., J. R. McCabe, et al. (1999). "The Drosophila beta FTZ-F1 orphan nuclear receptor provides competence for stage-specific responses to the steroid hormone ecdysone." Mol Cell **3**(2): 143-9.

Brown, V., P. Jin, et al. (2001). "Microarray identification of FMRP-associated brain mRNAs and altered mRNA translational profiles in fragile X syndrome." Cell **107**(4): 477-87.

Burchard, J., A. L. Jackson, et al. (2009). "MicroRNA-like off-target transcript regulation by siRNAs is species specific." RNA **15**(2): 308-15.

Cai, X., S. Lu, et al. (2005). "Kaposi's sarcoma-associated herpesvirus expresses an array of viral microRNAs in latently infected cells." Proc Natl Acad Sci U S A **102**(15): 5570-5.

Calin, G. A., C. D. Dumitru, et al. (2002). "Frequent deletions and down-regulation of micro- RNA genes miR15 and miR16 at 13q14 in chronic lymphocytic leukemia." Proc Natl Acad Sci U S A **99**(24): 15524-9.

Calin, G. A., C. Sevignani, et al. (2004). "Human microRNA genes are frequently located at fragile sites and genomic regions involved in cancers." Proc Natl Acad Sci U S A **101**(9): 2999-3004.

Castanotto, D., K. Sakurai, et al. (2007). "Combinatorial delivery of small interfering RNAs reduces RNAi efficacy by selective incorporation into RISC." Nucleic Acids Res **35**(15): 5154-64.

Caudy, A. A., M. Myers, et al. (2002). "Fragile X-related protein and VIG associate with the RNA interference machinery." Genes Dev **16**(19): 2491-6.

Ceman, S., V. Brown, et al. (1999). "Isolation of an FMRP-associated messenger ribonucleoprotein particle and identification of nucleolin and the fragile X-related proteins as components of the complex." Mol Cell Biol **19**(12): 7925-32.

Cheloufi, S., C. O. Dos Santos, et al. "A dicer-independent miRNA biogenesis pathway that requires Ago catalysis." Nature.

Chen, J. F., E. M. Mandel, et al. (2006). "The role of microRNA-1 and microR-NA-133 in skeletal muscle proliferation and differentiation." Nat Genet **38**(2): 228-33.

Chen, L., S. W. Yun, et al. (2003). "The fragile X mental retardation protein binds and regulates a novel class of mRNAs containing U rich target sequences." Neuroscience **120**(4): 1005-17.

Chen, X. (2004). "A microRNA as a translational repressor of APETALA2 in Arabidopsis flower development." Science **303**(5666): 2022-5.

Chisholm, A. and M. Tessier-Lavigne (1999). "Conservation and divergence of axon guidance mechanisms." Curr Opin Neurobiol **9**(5): 603-15.

Cifuentes, D., H. Xue, et al. "A Novel miRNA Processing Pathway Independent of Dicer Requires Argonaute2 Catalytic Activity." Science.

Comet, J. P. and J. Henry (2002). "Pairwise sequence alignment using a PROS-ITE pattern-derived similarity score." Comput Chem **26**(5): 421-36.

Corsten, M. F., R. Miranda, et al. (2007). "MicroRNA-21 knockdown disrupts glioma growth in vivo and displays synergistic cytotoxicity with neural precursor cell delivered S-TRAIL in human gliomas." Cancer Res **67**(19): 8994-9000.

Creighton, C. J., A. L. Benham, et al. "Discovery of novel microRNAs in female reproductive tract using next generation sequencing." PLoS One **5**(3): e9637.

Cullen, B. R. (2006). "Viruses and microRNAs." Nat Genet **38 Suppl**: S25-30.

Cummins, J. M., Y. He, et al. (2006). "The colorectal microRNAome." Proc Natl Acad Sci U S A **103**(10): 3687-92.

D'Avino, P. P., S. Crispi, et al. (1995). "The moulting hormone ecdysone is able to recognize target elements composed of direct repeats." Mol Cell Endocrinol **113**(1): 1-9.

Darnell, J. C., K. B. Jensen, et al. (2001). "Fragile X mental retardation protein targets G quartet mRNAs important for neuronal function." Cell **107**(4): 489-99.

Davidson, E. H. and R. J. Britten (1979). "Regulation of gene expression: possible role of repetitive sequences." Science **204**(4397): 1052-9.

Denman, R. B. (2003). "Deja vu all over again: FMRP binds U-rich target mRNAs." Biochem Biophys Res Commun **310**(1): 1-7.

Didiano, D. and O. Hobert (2006). "Perfect seed pairing is not a generally reliable predictor for miRNA-target interactions." Nat Struct Mol Biol **13**(9): 849-51.

Didiano, D. and O. Hobert (2008). "Molecular architecture of a miRNA-regulated 3' UTR." RNA **14**(7): 1297-317.

Doench, J. G., C. P. Petersen, et al. (2003). "siRNAs can function as miRNAs." Genes Dev **17**(4): 438-42.

Doench, J. G. and P. A. Sharp (2004). "Specificity of microRNA target selection in translational repression." Genes Dev **18**(5): 504-11.

Dolzhanskaya, N., Y. J. Sung, et al. (2003). "The fragile X mental retardation protein interacts with U-rich RNAs in a yeast three-hybrid system." Biochem Biophys Res Commun **305**(2): 434-41.

Ebert, M., J. Neilson, et al. (2007). "MicroRNA sponges: competitive inhibitors of small RNAs in mammalian cells." Nat Methods **4**(9): 721-726.

Eis, P. S., W. Tam, et al. (2005). "Accumulation of miR-155 and BIC RNA in human B cell lymphomas." Proc Natl Acad Sci U S A **102**(10): 3627-32.

Elbashir, S. M., J. Martinez, et al. (2001). "Functional anatomy of siRNAs for mediating efficient RNAi in Drosophila melanogaster embryo lysate." EMBO J **20**(23): 6877-88.

Elmen, J., M. Lindow, et al. (2008). "LNA-mediated microRNA silencing in non-human primates." Nature **452**(7189): 896-9.

Elmen, J., M. Lindow, et al. (2008). "Antagonism of microRNA-122 in mice by systemically administered LNA-antimiR leads to up-regulation of a large set of predicted target mRNAs in the liver." Nucleic Acids Res **36**(4): 1153-62.

Emde, A. K., M. Grunert, et al. "MicroRazerS: rapid alignment of small RNA reads." Bioinformatics **26**(1): 123-4.

Enright, A., B. John, et al. (2003). "MicroRNA targets in Drosophila." Genome Biol **5**(1): R1.

Enright, A. J., B. John, et al. (2003). "MicroRNA targets in Drosophila." Genome Biol **5**(1): R1.

Esau, C., S. Davis, et al. (2006). "miR-122 regulation of lipid metabolism revealed by in vivo antisense targeting." Cell Metab **3**(2): 87-98.

Esquela-Kerscher, A., P. Trang, et al. (2008). "The let-7 microRNA reduces tumor growth in mouse models of lung cancer." Cell Cycle **7**(6): 759-64.

Fabani, M. M. and M. J. Gait (2008). "miR-122 targeting with LNA/2'-O-methyl oligonucleotide mixmers, peptide nucleic acids (PNA), and PNA-peptide conjugates." Rna **14**(2): 336-46.

Farh, K., A. Grimson, et al. (2005). "The widespread impact of mammalian MicroRNAs on mRNA repression and evolution." Science **310**(5755): 1817-21.

Filipowicz, W., S. Bhattacharyya, et al. (2008). "Mechanisms of post-transcriptional regulation by microRNAs: are the answers in sight?" Nat Rev Genet **9**(2): 102-14.

Filipowicz, W., S. N. Bhattacharyya, et al. (2008). "Mechanisms of post-transcriptional regulation by microRNAs: are the answers in sight?" Nat Rev Genet **9**(2): 102-14.

Flanagan, J., S. Healey, et al. (2003). "Analysis of the transcription regulator, CNOT7, as a candidate chromosome 8 tumor suppressor gene in colorectal cancer." Int J Cancer **106**(4): 505-9.

Forbes, S. A., G. Bhamra, et al. (2008). "The Catalogue of Somatic Mutations in Cancer (COSMIC)." Curr Protoc Hum Genet **Chapter 10**: Unit 10 11.

Franco-Zorrilla, J. M., A. Valli, et al. (2007). "Target mimicry provides a new mechanism for regulation of microRNA activity." Nat Genet **39**(8): 1033-1037.

Friedlander, M. R., W. Chen, et al. (2008). "Discovering microRNAs from deep sequencing data using miRDeep." Nat Biotechnol **26**(4): 407-15.

Friedman, R. C., K. K. Farh, et al. (2009). "Most mammalian mRNAs are conserved targets of microRNAs." Genome Res **19**(1): 92-105.

Futreal, P. A., L. Coin, et al. (2004). "A census of human cancer genes." Nat Rev Cancer **4**(3): 177-83.

Ganem, D. and J. Ziegelbauer (2008). "MicroRNAs of Kaposi's sarcoma-associated herpes virus." Semin Cancer Biol **18**(6): 437-40.

Garber, K., K. T. Smith, et al. (2006). "Transcription, translation and fragile X syndrome." Curr Opin Genet Dev **16**(3): 270-5.

Gauthier, N. P., M. E. Larsen, et al. (2008). "Cyclebase.org--a comprehensive multi-organism online database of cell-cycle experiments." Nucleic Acids Res **36** (Database issue): D854-9.

Ghildiyal, M. and P. D. Zamore (2009). "Small silencing RNAs: an expanding universe." Nat Rev Genet **10**(2): 94-108.

Giniger, E. (2002). "How do Rho family GTPases direct axon growth and guidance? A proposal relating signaling pathways to growth cone mechanics." Differentiation **70**(8): 385-96.

Giraldez, A. J., R. M. Cinalli, et al. (2005). "MicroRNAs regulate brain morphogenesis in zebrafish." Science **308**(5723): 833-8.

Gottwein, E., N. Mukherjee, et al. (2007). "A viral microRNA functions as an orthologue of cellular miR-155." Nature **450**(7172): 1096-9.

Griffiths-Jones, S. (2004). "The microRNA Registry." Nucleic Acids Res **32**(Database issue): D109-11.

Grimm, D., K. L. Streetz, et al. (2006). "Fatality in mice due to oversaturation of cellular microRNA/short hairpin RNA pathways." Nature **441**(7092): 537-41.

Grimson, A., K. K. Farh, et al. (2007). "MicroRNA targeting specificity in mammals: determinants beyond seed pairing." Mol Cell **27**(1): 91-105.

Grimson, A., K. K.-H. Farh, et al. (2007). "MicroRNA Targeting Specificity in Mammals: Determinants beyond Seed Pairing." Molecular cell **27**(1): 91-105.

Gruber, J., T. Lampe, et al. (2005). "RNAi of FACE1 protease results in growth inhibition of human cells expressing lamin A: implications for Hutchinson-Gilford progeria syndrome." J Cell Sci **118**(Pt 4): 689-96.

Hafner, M., P. Landgraf, et al. (2008). "Identification of microRNAs and other small regulatory RNAs using cDNA library sequencing." Methods **44**(1): 3-12.

Hafner, M., M. Landthaler, et al. "Transcriptome-wide identification of RNA-binding protein and microRNA target sites by PAR-CLIP." Cell **141**(1): 129-41.

Hake, L. E., R. Mendez, et al. (1998). "Specificity of RNA binding by CPEB: requirement for RNA recognition motifs and a novel zinc finger." Mol Cell Biol **18**(2): 685-93.

Haley, B. and P. Zamore (2004). "Kinetic analysis of the RNAi enzyme complex." Nat Struct Mol Biol **11**(7): 599.

Hammell, M. "Computational methods to identify miRNA targets." Semin Cell Dev Biol.

Hammell, M., D. Long, et al. (2008). "mirWIP: microRNA target prediction based on microRNA-containing ribonucleoprotein-enriched transcripts." Nature Methods **5**(9): 813-819.

Hammell, M., D. Long, et al. (2008). "mirWIP: microRNA target prediction based on microRNA-containing ribonucleoprotein-enriched transcripts." Nat Methods **5**(9): 813-9.

He, L. and G. J. Hannon (2004). "MicroRNAs: small RNAs with a big role in gene regulation." Nat Rev Genet **5**(7): 522-31.

He, L., X. He, et al. (2007). "A microRNA component of the p53 tumour suppressor network." Nature **447**(7148): 1130-4.

He, L., X. He, et al. (2007). "A microRNA component of the p53 tumour suppressor network." Nature **447**(7148): 1130-4.

Hicks, J. A., N. Trakooljul, et al. "Discovery of chicken microRNAs associated with lipogenesis and cell proliferation." Physiol Genomics.

Hofacker, I. L. and P. F. Stadler (2006). "Memory efficient folding algorithms for circular RNA secondary structures." Bioinformatics **22**(10): 1172-6.

Houbaviy, H. B., M. F. Murray, et al. (2003). "Embryonic stem cell-specific MicroRNAs." Dev Cell **5**(2): 351-8.

Huang, Y. S., J. H. Carson, et al. (2003). "Facilitation of dendritic mRNA transport by CPEB." Genes Dev **17**(5): 638-53.

Hubbard, T., D. Barker, et al. (2002). "The Ensembl genome database project." Nucleic Acids Res **30**(1): 38-41.

Hutvagner, G. and P. D. Zamore (2002). "A microRNA in a multiple-turnover RNAi enzyme complex." Science **297**(5589): 2056-60.

Jackson, A. L., S. R. Bartz, et al. (2003). "Expression profiling reveals off-target gene regulation by RNAi." Nat Biotechnol **21**(6): 635-7.

Jackson, A. L., J. Burchard, et al. (2006). "Position-specific chemical modification of siRNAs reduces "off-target" transcript silencing." RNA **12**(7): 1197-205.

Jackson, A. L., J. Burchard, et al. (2006). "Widespread siRNA "off-target" transcript silencing mediated by seed region sequence complementarity." RNA (New York, N.Y.) **12**(7): 1179-1187.

Jackson, A. L., J. Burchard, et al. (2006). "Widespread siRNA "off-target" transcript silencing mediated by seed region sequence complementarity." Rna **12**(7): 1179-87.

Jacob, F. and J. Monod (1961). "Genetic regulatory mechanisms in the synthesis of proteins." J Mol Biol **3**: 318-56.

Jiang, C., E. H. Baehrecke, et al. (1997). "Steroid regulated programmed cell death during Drosophila metamorphosis." Development **124**(22): 4673-83.

Jiang, C., A. F. Lamblin, et al. (2000). "A steroid-triggered transcriptional hierarchy controls salivary gland cell death during Drosophila metamorphosis." Mol Cell **5**(3): 445-55.

Jin, P., R. S. Alisch, et al. (2004). "RNA and microRNAs in fragile X mental retardation." Nat Cell Biol **6**(11): 1048-53.

Jin, P., D. C. Zarnescu, et al. (2004). "Biochemical and genetic interaction between the fragile X mental retardation protein and the microRNA pathway." Nat Neurosci **7**(2): 113-7.

John, B., A. J. Enright, et al. (2004). "Human MicroRNA targets." PLoS Biol **2**(11): e363.

John, B., C. Sander, et al. (2006). "Prediction of human microRNA targets." Methods Mol Biol **342**: 101-13.

John, M., R. Constien, et al. (2007). "Effective RNAi-mediated gene silencing without interruption of the endogenous microRNA pathway." Nature **449**(7163): 745-7.

Johnston, R. J. and O. Hobert (2003). "A microRNA controlling left/right neuronal asymmetry in Caenorhabditis elegans." Nature **426**(6968): 845-9.

Jopling, C. L., M. Yi, et al. (2005). "Modulation of hepatitis C virus RNA abundance by a liver-specific MicroRNA." Science **309**(5740): 1577-81.

Kadesch, T. (2000). "Notch signaling: a dance of proteins changing partners." Exp Cell Res **260**(1): 1-8.

Kaminker, J. S., C. M. Bergman, et al. (2002). "The transposable elements of the Drosophila melanogaster euchromatin: a genomics perspective." Genome Biol **3**(12): RESEARCH0084.

Kertesz, M., N. Iovino, et al. (2007). "The role of site accessibility in microRNA target recognition." Nat Genet **39**(10): 1278-84.

Khan, A., D. Betel, et al. (2009). "Transfection of small RNAs globally perturbs gene regulation by endogenous microRNAs." Nat Biotech **27**(6): 549-555.

Khan, A. A., D. Betel, et al. (2009). "Transfection of small RNAs globally perturbs gene regulation by endogenous microRNAs." Nat Biotechnol **27**(6): 549-55.

Khvorova, A., A. Reynolds, et al. (2003). "Functional siRNAs and miRNAs exhibit strand bias." Cell **115**(2): 209-16.

Kim, J., A. Krichevsky, et al. (2004). "Identification of many microRNAs that co-purify with polyribosomes in mammalian neurons." Proc Natl Acad Sci U S A **101**(1): 360-5.

Kim, S. K., J. W. Nam, et al. (2006). "miTarget: microRNA target gene prediction using a support vector machine." BMC Bioinformatics **7**: 411.

Kiriakidou, M., P. T. Nelson, et al. (2004). "A combined computational-experimental approach predicts human microRNA targets." Genes Dev **18**(10): 1165-78.

Kittler, R., V. Surendranath, et al. (2007). "Genome-wide resources of endoribonuclease-prepared short interfering RNAs for specific loss-of-function studies." Nat Methods **4**(4): 337-44.

Koh, W., C. T. Sheng, et al. "Analysis of deep sequencing microRNA expression profile from human embryonic stem cells derived mesenchymal stem cells reveals possible role of let-7 microRNA family in downstream targeting of hepatic nuclear factor 4 alpha." BMC Genomics **11 Suppl 1**: S6.

Krek, A., D. Grun, et al. (2005). "Combinatorial microRNA target predictions." Nat Genet **37**(5): 495-500.

Krichevsky, A. M., K. S. King, et al. (2003). "A microRNA array reveals extensive regulation of microRNAs during brain development." RNA **9**(10): 1274-81.

Krutzfeldt, J., N. Rajewsky, et al. (2005). "Silencing of microRNAs in vivo with 'antagomirs'." Nature **438**(7068): 685-9.

Kumar, M. S., J. Lu, et al. (2007). "Impaired microRNA processing enhances cellular transformation and tumorigenesis." Nat Genet **39**(5): 673-7.

Lagos-Quintana, M., R. Rauhut, et al. (2001). "Identification of novel genes coding for small expressed RNAs." Science **294**(5543): 853-8.

Lagos-Quintana, M., R. Rauhut, et al. (2003). "New microRNAs from mouse and human." RNA **9**(2): 175-9.

Lagos-Quintana, M., R. Rauhut, et al. (2002). "Identification of tissue-specific microRNAs from mouse." Curr Biol **12**(9): 735-9.

Lai, E. C. and J. W. Posakony (1997). "The Bearded box, a novel 3' UTR sequence motif, mediates negative post-transcriptional regulation of Bearded and Enhancer of split Complex gene expression." Development **124**(23): 4847-56.

Lai, E. C., P. Tomancak, et al. (2003). "Computational identification of Drosophila microRNA genes." Genome Biol **4**(7): R42.

Landgraf, P., M. Rusu, et al. (2007). "A mammalian microRNA expression atlas based on small RNA library sequencing." Cell **129**(7): 1401-14.

Landthaler, M., D. Gaidatzis, et al. (2008). "Molecular characterization of human Argonaute-containing ribonucleoprotein complexes and their bound target mRNAs." Rna **14**(12): 2580-96.

Lau, N. C., L. P. Lim, et al. (2001). "An abundant class of tiny RNAs with probable regulatory roles in Caenorhabditis elegans." Science **294**(5543): 858-62.

Leachman, S. A., R. P. Hickerson, et al. (2008). "Therapeutic siRNAs for dominant genetic skin disorders including pachyonychia congenita." J Dermatol Sci **51**(3): 151-7.

Leaman, D., P. Y. Chen, et al. (2005). "Antisense-mediated depletion reveals essential and specific functions of microRNAs in Drosophila development." Cell **121**(7): 1097-108.

Lee, R. C. and V. Ambros (2001). "An extensive class of small RNAs in Caenorhabditis elegans." Science **294**(5543): 862-4.

Lee, R. C., R. L. Feinbaum, et al. (1993). "The C. elegans heterochronic gene lin-4 encodes small RNAs with antisense complementarity to lin-14." Cell **75**(5): 843-54.

Lewis, B., C. Burge, et al. (2005). "Conserved seed pairing, often flanked by adenosines, indicates that thousands of human genes are microRNA targets." Cell **120**(1): 15-20.

Lewis, B., I.-h. Shih, et al. (2003). "Prediction of Mammalian MicroRNA Targets." Cell **115**(7): 787-798.

Lewis, B. P., C. B. Burge, et al. (2005). "Conserved seed pairing, often flanked by adenosines, indicates that thousands of human genes are microRNA targets." Cell **120**(1): 15-20.

Lewis, B. P., I. H. Shih, et al. (2003). "Prediction of mammalian microRNA targets." Cell **115**(7): 787-98.

Lim, L., N. Lau, et al. (2005). "Microarray analysis shows that some microRNAs downregulate large numbers of target mRNAs." Nature **433**(7027): 769-773.

Lim, L. P., N. C. Lau, et al. (2005). "Microarray analysis shows that some microRNAs downregulate large numbers of target mRNAs." Nature **433**(7027): 769-73.

Lim, L. P., N. C. Lau, et al. (2003). "The microRNAs of Caenorhabditis elegans." Genes Dev **17**(8): 991-1008.

Linsley, P., J. Schelter, et al. (2007). "Transcripts targeted by the microRNA-16 family cooperatively regulate cell cycle progression." Molecular and Cellular Biology **27**(6): 2240-2252.

Linsley, P. S., J. Schelter, et al. (2007). "Transcripts targeted by the microRNA-16 family cooperatively regulate cell cycle progression." Mol Cell Biol **27**(6): 2240-52.

Llave, C., Z. Xie, et al. (2002). "Cleavage of Scarecrow-like mRNA targets directed by a class of Arabidopsis miRNA." Science **297**(5589): 2053-6.

Lohe, A. R., A. J. Hilliker, et al. (1993). "Mapping simple repeated DNA sequences in heterochromatin of Drosophila melanogaster." Genetics **134**(4): 1149-74.

Lu, J., Y. Fu, et al. (2008). "Adaptive evolution of newly emerged micro-RNA genes in Drosophila." Mol Biol Evol **25**(5): 929-38.

Lu, J., G. Getz, et al. (2005). "MicroRNA expression profiles classify human cancers." Nature **435**(7043): 834-8.

Lu, J., Y. Shen, et al. (2008). "The birth and death of microRNA genes in Drosophila." Nat Genet **40**(3): 351-5.

Lupberger, J., L. Brino, et al. (2008). "RNAi: a powerful tool to unravel hepatitis C virus-host interactions within the infectious life cycle." J Hepatol **48**(3): 523-5.

Ma, J. B., K. Ye, et al. (2004). "Structural basis for overhang-specific small interfering RNA recognition by the PAZ domain." Nature **429**(6989): 318-22.

Ma, L., J. Teruya-Feldstein, et al. (2007). "Tumour invasion and metastasis initiated by microRNA-10b in breast cancer." Nature **449**(7163): 682-8.

Mann, R. S. (1997). "Why are Hox genes clustered?" Bioessays **19**(8): 661-4.

Margolin, A. A., S. E. Ong, et al. (2009). "Empirical Bayes analysis of quantitative proteomics experiments." PLoS One **4**(10): e7454.

Mattick, J. S. and M. J. Gagen (2001). "The evolution of controlled multitasked gene networks: the role of introns and other noncoding RNAs in the development of complex organisms." Mol Biol Evol **18**(9): 1611-30.

Mayr, C., M. T. Hemann, et al. (2007). "Disrupting the pairing between let-7 and Hmga2 enhances oncogenic transformation." Science **315**(5818): 1576-9.

McBride, J. L., R. L. Boudreau, et al. (2008). "Artificial miRNAs mitigate shRNA-mediated toxicity in the brain: implications for the therapeutic development of RNAi." Proc Natl Acad Sci U S A **105**(15): 5868-73.

McGinnis, N., M. A. Kuziora, et al. (1990). "Human Hox-4.2 and Drosophila deformed encode similar regulatory specificities in Drosophila embryos and larvae." Cell **63**(5): 969-76.

Mendez, R., D. Barnard, et al. (2002). "Differential mRNA translation and meiotic progression require Cdc2-mediated CPEB destruction." EMBO J **21**(7): 1833-44.

Mendez, R., L. E. Hake, et al. (2000). "Phosphorylation of CPE binding factor by Eg2 regulates translation of c-mos mRNA." Nature **404**(6775): 302-7.

Mendez, R., K. G. Murthy, et al. (2000). "Phosphorylation of CPEB by Eg2 mediates the recruitment of CPSF into an active cytoplasmic polyadenylation complex." Mol Cell **6**(5): 1253-9.

Mendez, R. and J. D. Richter (2001). "Translational control by CPEB: a means to the end." Nat Rev Mol Cell Biol **2**(7): 521-9.

Metzler, M., M. Wilda, et al. (2004). "High expression of precursor microRNA-155/BIC RNA in children with Burkitt lymphoma." Genes Chromosomes Cancer **39**(2): 167-9.

Michael, M. Z., O. C. SM, et al. (2003). "Reduced accumulation of specific microRNAs in colorectal neoplasia." Mol Cancer Res **1**(12): 882-91.

Minakhina, S., J. Yang, et al. (2003). "Tamo selectively modulates nuclear import in Drosophila." Genes Cells **8**(4): 299-310.

Miranda, K. C., T. Huynh, et al. (2006). "A pattern-based method for the identification of MicroRNA binding sites and their corresponding heteroduplexes." Cell **126**(6): 1203-17.

Miyashiro, K. Y., A. Beckel-Mitchener, et al. (2003). "RNA cargoes associating with FMRP reveal deficits in cellular functioning in Fmr1 null mice." Neuron 37(3): 417-31.

Morin, R., M. Bainbridge, et al. (2008). "Profiling the HeLa S3 transcriptome using randomly primed cDNA and massively parallel short-read sequencing." BioTechniques 45(1): 81-94.

Mortazavi, A., B. Williams, et al. (2008). "Mapping and quantifying mammalian transcriptomes by RNA-Seq." Nature methods 5(7): 621-628.

Moss, E. G., R. C. Lee, et al. (1997). "The cold shock domain protein LIN-28 controls developmental timing in C. elegans and is regulated by the lin-4 RNA." Cell 88(5): 637-46.

Moss, E. G. and L. Tang (2003). "Conservation of the heterochronic regulator Lin-28, its developmental expression and microRNA complementary sites." Dev Biol 258(2): 432-42.

Mu, P., Y. C. Han, et al. (2009). "Genetic dissection of the miR-17~92 cluster of microRNAs in Myc-induced B-cell lymphomas." Genes Dev 23(24): 2806-11.

Mulder, N. J., R. Apweiler, et al. (2003). "The InterPro Database, 2003 brings increased coverage and new features." Nucleic Acids Res 31(1): 315-8.

Muller, S., A. Ledl, et al. (2004). "SUMO: a regulator of gene expression and genome integrity." Oncogene 23(11): 1998-2008.

Murchison, E. P., C. Tovar, et al. "The Tasmanian devil transcriptome reveals Schwann cell origins of a clonally transmissible cancer." Science 327(5961): 84-7.

Nagashima, M., M. Shiseki, et al. (2001). "DNA damage-inducible gene p33ING2 negatively regulates cell proliferation through acetylation of p53." Proc Natl Acad Sci U S A 98(17): 9671-6.

Needleman, S. B. and C. D. Wunsch (1970). "A general method applicable to the search for similarities in the amino acid sequence of two proteins." J Mol Biol 48(3): 443-53.

Nielsen, C. B., N. Shomron, et al. (2007). "Determinants of targeting by endogenous and exogenous microRNAs and siRNAs." Rna 13(11): 1894-910.

Oates, A. C., A. E. Bruce, et al. (2000). "Too much interference: injection of double-stranded RNA has nonspecific effects in the zebrafish embryo." Dev Biol 224(1): 20-8.

Pal-Bhadra, M., U. Bhadra, et al. (2002). "RNAi related mechanisms affect both transcriptional and posttranscriptional transgene silencing in Drosophila." Mol Cell **9**(2): 315-27.

Palatnik, J. F., E. Allen, et al. (2003). "Control of leaf morphogenesis by microR-NAs." Nature **425**(6955): 257-63.

Pan, H., W. X. Qin, et al. (2001). "Cloning, mapping, and characterization of a human homologue of the yeast longevity assurance gene LAG1." Genomics **77**(1-2): 58-64.

Pasquinelli, A. E., B. J. Reinhart, et al. (2000). "Conservation of the sequence and temporal expression of let-7 heterochronic regulatory RNA." Nature **408**(6808): 86-9.

Pesole, G., S. Liuni, et al. (2002). "UTRdb and UTRsite: specialized databases of sequences and functional elements of 5' and 3' untranslated regions of eukaryotic mRNAs. Update 2002." Nucleic Acids Res **30**(1): 335-40.

Pfeffer, S., A. Sewer, et al. (2005). "Identification of microRNAs of the herpesvirus family." Nat Methods **2**(4): 269-76.

Pfeffer, S., M. Zavolan, et al. (2004). "Identification of virus-encoded microR-NAs." Science **304**(5671): 734-6.

Plante, I., L. Davidovic, et al. (2006). "Dicer-Derived MicroRNAs Are Utilized by the Fragile X Mental Retardation Protein for Assembly on Target RNAs." J Biomed Biotechnol **2006**(4): 64347.

Plante, I. and P. Provost (2006). "Hypothesis: A Role for Fragile X Mental Retardation Protein in Mediating and Relieving MicroRNA-Guided Translational Repression?" J Biomed Biotechnol **2006**(4): 16806.

Rajagopalan, L. E. and J. S. Malter (2000). "Growth factor-mediated stabilization of amyloid precursor protein mRNA is mediated by a conserved 29-nucleotide sequence in the 3'-untranslated region." J Neurochem **74**(1): 52-9.

Rajasethupathy, P., F. Fiumara, et al. (2009). "Characterization of small RNAs in aplysia reveals a role for miR-124 in constraining synaptic plasticity through CREB." Neuron **63**(6): 803-17.

Rajewsky, N. and N. D. Socci (2004). "Computational identification of microRNA targets." Dev Biol **267**(2): 529-35.

Ramos, A., D. Hollingworth, et al. (2003). "G-quartet-dependent recognition between the FMRP RGG box and RNA." RNA **9**(10): 1198-207.

Ramsköld, D., E. Wang, et al. (2009). "An abundance of ubiquitously expressed genes revealed by tissue transcriptome sequence data." PLoS Comput Biol **5**(12): e1000598.

Rayburn, L. Y., H. C. Gooding, et al. (2003). "amontillado, the Drosophila homolog of the prohormone processing protease PC2, is required during embryogenesis and early larval development." Genetics **163**(1): 227-37.

Rehmsmeier, M., P. Steffen, et al. (2004). "Fast and effective prediction of microRNA/target duplexes." RNA **10**(10): 1507-17.

Reinhart, B. J. and G. Ruvkun (2001). "Isoform-specific mutations in the Caenorhabditis elegans heterochronic gene lin-14 affect stage-specific patterning." Genetics **157**(1): 199-209.

Reinhart, B. J., F. J. Slack, et al. (2000). "The 21-nucleotide let-7 RNA regulates developmental timing in Caenorhabditis elegans." Nature **403**(6772): 901-6.

Reynolds, A., D. Leake, et al. (2004). "Rational siRNA design for RNA interference." Nat Biotechnol **22**(3): 326-30.

Rhoades, M. W., B. J. Reinhart, et al. (2002). "Prediction of plant microRNA targets." Cell **110**(4): 513-20.

Riddiford, L. M., P. Cherbas, et al. (2000). "Ecdysone receptors and their biological actions." Vitam Horm **60**: 1-73.

Rieckhof, G. E., F. Casares, et al. (1997). "Nuclear translocation of extradenticle requires homothorax, which encodes an extradenticle-related homeodomain protein." Cell **91**(2): 171-83.

Ritchie, W., S. Flamant, et al. (2009). "Predicting microRNA targets and functions: traps for the unwary." Nat Methods **6**(6): 397-8.

Robbins, M., A. Judge, et al. (2008). "Misinterpreting the therapeutic effects of siRNA caused by immune stimulation." Hum Gene Ther.

Rossi, J., P. Zamore, et al. (2008). "Wandering eye for RNAi." Nat Med **14**(6): 611.

Rusinov, V., V. Baev, et al. (2005). "MicroInspector: a web tool for detection of miRNA binding sites in an RNA sequence." Nucleic Acids Res **33**(Web Server issue): W696-700.

Ruvkun, G. (2008). "The perfect storm of tiny RNAs." Nat Med **14**(10): 1041-5.

Rybak, A., H. Fuchs, et al. (2008). "A feedback loop comprising lin-28 and let-7 controls pre-let-7 maturation during neural stem-cell commitment." Nat Cell Biol **10**(8): 987-93.

Saetrom, P., B. S. Heale, et al. (2007). "Distance constraints between microRNA target sites dictate efficacy and cooperativity." Nucleic Acids Res **35**(7): 2333-42.

Sandberg, R., J. R. Neilson, et al. (2008). "Proliferating cells express mRNAs with shortened 3' untranslated regions and fewer microRNA target sites." Science **320**(5883): 1643-7.

Schwarz, D. S., H. Ding, et al. (2006). "Designing siRNA That Distinguish between Genes That Differ by a Single Nucleotide." PLoS Genetics **2**(9): e140.

Seitz, H. and P. D. Zamore (2006). "Rethinking the microprocessor." Cell **125**(5): 827-9.

Selbach, M., B. Schwanhausser, et al. (2008). "Widespread changes in protein synthesis induced by microRNAs." Nature.

Selbach, M., B. Schwanhäusser, et al. (2008). "Widespread changes in protein synthesis induced by microRNAs." Nature **455**(7209): 58.

Sempere, L. F., S. Freemantle, et al. (2004). "Expression profiling of mammalian microRNAs uncovers a subset of brain-expressed microRNAs with possible roles in murine and human neuronal differentiation." Genome Biol **5**(3): R13.

Sempere, L. F., N. S. Sokol, et al. (2003). "Temporal regulation of microRNA expression in Drosophila melanogaster mediated by hormonal signals and broad-Complex gene activity." Dev Biol **259**(1): 9-18.

Sethupathy, P., M. Megraw, et al. (2006). "A guide through present computational approaches for the identification of mammalian microRNA targets." Nat Methods **3**(11): 881-6.

Shibata, M. A., J. Morimoto, et al. (2008). "Combination therapy with short interfering RNA vectors against VEGF-C and VEGF-A suppresses lymph node and lung metastasis in a mouse immunocompetent mammary cancer model." Cancer Gene Ther **15**(12): 776-86.

Simon, A. F., C. Shih, et al. (2003). "Steroid control of longevity in Drosophila melanogaster." Science **299**(5611): 1407-10.

Simon, J. A. and J. W. Tamkun (2002). "Programming off and on states in chromatin: mechanisms of Polycomb and trithorax group complexes." Curr Opin Genet Dev **12**(2): 210-8.

Smith, T. F. and M. S. Waterman (1981). "Identification of common molecular subsequences." J Mol Biol **147**(1): 195-7.

Smith, T. F., M. S. Waterman, et al. (1981). "Comparative biosequence metrics." J Mol Evol **18**(1): 38-46.

Stark, A., J. Brennecke, et al. (2005). "Animal MicroRNAs confer robustness to gene expression and have a significant impact on 3'UTR evolution." Cell **123**(6): 1133-46.

Stark, A., J. Brennecke, et al. (2003). "Identification of Drosophila MicroRNA targets." PLoS Biol **1**(3): E60.

Stark, A., N. Bushati, et al. (2008). "A single Hox locus in Drosophila produces functional microRNAs from opposite DNA strands." Genes Dev **22**(1): 8-13.

Stark, M. S., S. Tyagi, et al. "Characterization of the Melanoma miRNAome by Deep Sequencing." PLoS One **5**(3): e9685.

Steitz, J. A. and S. Vasudevan (2009). "miRNPs: versatile regulators of gene expression in vertebrate cells." Biochem Soc Trans **37**(Pt 5): 931-5.

Stenvang, J., M. Lindow, et al. (2008). "Targeting of microRNAs for therapeutics." Biochem Soc Trans **36**(Pt 6): 1197-200.

Steward, O. and E. M. Schuman (2003). "Compartmentalized synthesis and degradation of proteins in neurons." Neuron **40**(2): 347-59.

Stewart, C. K., J. Li, et al. (2008). "Adverse effects induced by short hairpin RNA expression in porcine fetal fibroblasts." Biochem Biophys Res Commun **370**(1): 113-7.

Stratton, M. R., P. J. Campbell, et al. (2009). "The cancer genome." Nature **458**(7239): 719-24.

Sun, G., H. Li, et al. (2010). "Sequence context outside the target region influences the effectiveness of miR-223 target sites in the RhoB 3'UTR." Nucleic Acids Res **38**(1): 239-52.

Szczyrba, J., E. Loprich, et al. "The microRNA profile of prostate carcinoma obtained by deep sequencing." Mol Cancer Res **8**(4): 529-38.

Tagami, Y., N. Inaba, et al. "Cloning new small RNA sequences." Methods Mol Biol **631**: 123-38.

Tam, W. (2008). "The emergent role of microRNAs in molecular diagnostics of cancer." J Mol Diagn **10**(5): 411-4.

Tang, G., B. J. Reinhart, et al. (2003). "A biochemical framework for RNA silencing in plants." Genes Dev **17**(1): 49-63.

Tapper, J., E. Kettunen, et al. (2001). "Changes in gene expression during progression of ovarian carcinoma." Cancer Genet Cytogenet **128**(1): 1-6.

Tavazoie, S. F., C. Alarcon, et al. (2008). "Endogenous human microRNAs that suppress breast cancer metastasis." Nature **451**(7175): 147-52.

Thierry-Mieg, D. and J. Thierry-Mieg (2006). "AceView: a comprehensive cDNA-supported gene and transcripts annotation." Genome Biology **7**(suppl 1): S12.

Thummel, C. S. (2001). "Molecular mechanisms of developmental timing in C. elegans and Drosophila." Dev Cell **1**(4): 453-65.

Thummel, C. S. (2001). "Steroid-triggered death by autophagy." Bioessays **23**(8): 677-82.

Todd, P. K., K. J. Mack, et al. (2003). "The fragile X mental retardation protein is required for type-I metabotropic glutamate receptor-dependent translation of PSD-95." Proc Natl Acad Sci U S A **100**(24): 14374-8.

Tomari, Y. and P. D. Zamore (2005). "MicroRNA biogenesis: drosha can't cut it without a partner." Curr Biol **15**(2): R61-4.

Umbach, J. L., M. F. Kramer, et al. (2008). "MicroRNAs expressed by herpes simplex virus 1 during latent infection regulate viral mRNAs." Nature **454**(7205): 780-3.

Umbach, J. L., M. A. Nagel, et al. (2009). "Analysis of human alphaherpesvirus microRNA expression in latently infected human trigeminal ganglia." J Virol **83**(20): 10677-83.

Umbach, J. L., K. Wang, et al. "Identification of viral microRNAs expressed in human sacral ganglia latently infected with herpes simplex virus 2." J Virol **84**(2): 1189-92.

Vankoningsloo, S., F. de Longueville, et al. (2008). "Gene expression silencing with 'specific' small interfering RNA goes beyond specificity - a study of key parameters to take into account in the onset of small interfering RNA off-target effects." Febs J **275**(11): 2738-53.

Vasudevan, S., Y. Tong, et al. (2007). "Switching from repression to activation: microRNAs can up-regulate translation." Science **318**(5858): 1931-4.

Vella, M. C., K. Reinert, et al. (2004). "Architecture of a validated microRNA::target interaction." Chem Biol **11**(12): 1619-23.

Ventura, A. and T. Jacks (2009). "MicroRNAs and cancer: short RNAs go a long way." Cell **136**(4): 586-91.

Waggoner, S. A. and S. A. Liebhaber (2003). "Identification of mRNAs associated with alphaCP2-containing RNP complexes." Mol Cell Biol **23**(19): 7055-67.

Wang, X. and X. Wang (2006). "Systematic identification of microRNA functions by combining target prediction and expression profiling." Nucleic Acids Res **34**(5): 1646-52.

Wargelius, A., S. Ellingsen, et al. (1999). "Double-stranded RNA induces specific developmental defects in zebrafish embryos." Biochem Biophys Res Commun **263**(1): 156-61.

Waterman, M. S. and M. Eggert (1987). "A new algorithm for best subsequence alignments with application to tRNA-rRNA comparisons." J Mol Biol **197**(4): 723-8.

Weiler, I. J., C. C. Spangler, et al. (2004). "Fragile X mental retardation protein is necessary for neurotransmitter-activated protein translation at synapses." Proc Natl Acad Sci U S A **101**(50): 17504-9.

Wheeler, B. M., A. M. Heimberg, et al. (2009). "The deep evolution of metazoan microRNAs." Evol Dev **11**(1): 50-68.

Wightman, B., I. Ha, et al. (1993). "Posttranscriptional regulation of the heterochronic gene lin-14 by lin-4 mediates temporal pattern formation in C. elegans." Cell **75**(5): 855-62.

Williams, R. W. and G. M. Rubin (2002). "ARGONAUTE1 is required for efficient RNA interference in Drosophila embryos." Proc Natl Acad Sci U S A **99**(10): 6889-94.

Wismar, J., T. Loffler, et al. (1995). "The Drosophila melanogaster tumor suppressor gene lethal(3)malignant brain tumor encodes a proline-rich protein with a novel zinc finger." Mech Dev **53**(1): 141-54.

Wuchty, S., W. Fontana, et al. (1999). "Complete suboptimal folding of RNA and the stability of secondary structures." Biopolymers **49**(2): 145-65.

Wyman, S. K., R. K. Parkin, et al. (2009). "Repertoire of microRNAs in epithelial ovarian cancer as determined by next generation sequencing of small RNA cDNA libraries." PLoS One **4**(4): e5311.

Xie, Z., L. K. Johansen, et al. (2004). "Genetic and functional diversification of small RNA pathways in plants." PLoS Biol **2**(5): E104.

Xu, P., S. Y. Vernooy, et al. (2003). "The Drosophila microRNA Mir-14 suppresses cell death and is required for normal fat metabolism." Curr Biol **13**(9): 790-5.

Yang, W., T. P. Chendrimada, et al. (2006). "Modulation of microRNA processing and expression through RNA editing by ADAR deaminases." Nat Struct Mol Biol **13**(1): 13-21.

Yang, Y., S. Xu, et al. (2009). "The bantam microRNA is associated with drosophila fragile X mental retardation protein and regulates the fate of germline stem cells." PLoS Genet **5**(4): e1000444.

Yart, A., P. Mayeux, et al. (2003). "Gab1, SHP-2 and other novel regulators of Ras: targets for anticancer drug discovery?" Curr Cancer Drug Targets **3**(3): 177-92.

Yokota, N., T. G. Mainprize, et al. (2004). "Identification of differentially expressed and developmentally regulated genes in medulloblastoma using suppression subtraction hybridization." Oncogene **23**(19): 3444-53.

Zisoulis, D. G., M. T. Lovci, et al. "Comprehensive discovery of endogenous Argonaute binding sites in Caenorhabditis elegans." Nat Struct Mol Biol **17**(2): 173-9.

Zuker, M. (2003). "Mfold web server for nucleic acid folding and hybridization prediction." Nucleic Acids Res **31**(13): 3406-15.

8   Curriculum Vitae

Debora Marks

Email: debbie@hms.harvard.edu

Telefon: ++ 1 617-953-4107

Adresse: 59 Evans Road, Brookline, MA 02445, USA

Staatsbürgerschaft: British Citizen / Britische Staatsbürgerin

Ausbildung

31.01.2007–aktuell. Institut für Theoretische Biologie, Humboldt Universität zu Berlin. Doktorandin

2008–aktuell. Forscherin. Systems Biology Department, Harvard Medical School, Boston.

2005–2008, wissenschaftliche Mitarbeiterin, Fontana Lab, Systems Biology Department, Boston. Alterungsprozesse in *C.elegans*. Genetik und molekularbiologisches Training

2002–2003 Spezialistin für Bioinformatik. Bauer Center for Genomics Research,Harvard University. Führung des Bioinformatikbereichs für das neue Zentrum, Anwerbung von Bauer Fellows. Unabhängige Forschung zu microRNAs

2000–2003 Spezialistin für Bioinformatik. Department of Cell Biology, Harvard Medical School, USA. Beratung der Fakultät zu bioinformatischer Forschung. Tutorin für Forschung in Bioinformatik

1998–2000 Post-doc-Stelle. Biology department, University of Manchester. Modelierung extrazellulärer Matrixproteine. Kryoelektronenmikroskopie von Collagen

1994–1998 Doktorandin. Wellcome Trust Prize Stipendium, Department of Pharmaceutical Sciences, University of Manchester, UK. Doktorarbeitstitel "Rational Drug Design in Parasitic Diseases". Proteinstrukturmodellierung (homology modeling) Ligandendesign, QSAR, Ligandendocking, NMR für chemisch modifizierte DNA.

1990–1994 BSc Hons Mathematics, University of Manchester, UK

1975–1980 2ndMBChB Medicine University of Bristol, UK

Forschungserfahrung

2008–2010   Systembiologie der microRNA-Regulation. (i) Sättigung in
RISC-Resultaten für die Deregulierung von Genen, die von endogenen
microRNAs reguliert werden. (ii) Wettkampf zwischen Zielmolekülen
um microRNAs schwächt mircoRNA/siRNAs-Effekte gegenüber indi-
viduellen Zielmolekülen (iii) Entdeckung von Hinweisen auf RNA-Bin-
deproteinen in Kooperation mit microRNA-Regulation, antagonistisch
und synergistisch (iv) mRNA-Halbwertszeit begrenzt RNAi-Wirkung (iv)
microRNAs regulieren 'dose-sensitive genes' in evolutionärem Rahmen.

2008–2010   Variation der Genomkopien in Säugetiergewebe. Entwurf
und Durchführung einer Studie zur Untersuchung der Varation der Genomkopien in verschiedenen Gewebearten des gleichen Tiers. 8 Mäu-
segewebe in 5 Mäusen wurden mithilfe von 244k Agilent Arrays und
maßgeschneiderter 1M probe Agilent Arrays auf Unterschiede in Genomkopien untersucht. Weiterführende qPCR-Untersuchungen von 10
Regionen in 6 Geweben.

2005–2008   Molekulare Grundlagen des Alterns. Alterungsstudien in
C. elegans mit dem Ziel, grundlegende molekulare Zustände zu iden-
tifizieren, welche die Lebensspanne vorhersagen. Entwicklung von
Reportermolekülen zur Messung von pH- und Redox-zustand in indi-
viduellen Wurmzellen. Diese zelltypspezifisch chemisch modifizierten
ratiometrischen GFP-Sensormoleküle, wurden in den Mitochondrien
dreier unterschiedlicher Zelltypen plaziert. Vorläufige Studien maßen
den mitochondrialen pH in Muskelzellen über die Lebensspanne mittels orangfluorezenter Tetramethylrosamine. Erste Ergebnisse deuten
auf monotone Änderungen des phs.

2001–2004   Vorhersage von microRNA-Funktionen. Entwicklung des
ersten microRNA-Zielvorhersage-Algorithmuses, Software und funk-
tioneller Analysen. Angewandt auf *Drosophila* melanogaster Genom
und (I) die kombinatorische Beziehung zwischen microRNAs und ihren
Zielgenen aufgedeckt und (ii) konservierte Zielgene zwischen anderen
Drosophila Spezies und Anopholeles gambia identifiziert. Identifikation
von microRNA-Regulation der Hox und Ecdyson Gene, die funktionelle
spezifischen microRNAs zugeordnet wurde, z.B. MiR-2 in Apoptose und

mir-7 in der Neurogenese.Entwicklung des miRanda-Algorithmus und Anwendung zur Vorhersage von humanen microRNA-Zielgenen. Viele Vorhersagen mittlerweile zahlreich in der experimentellen Literatur belegt, mehr als 750 Zitierungen der Publikation.

1994-2000 Erforschung von Proteinstruktueren, Entwicklung von Liganden-Kopplungs-Algorithmen, Einführung von Hochauflösungs-NMR-Spektroskopie zum Studium von chemisch modifizierten antisense Oligonukleotiden.

Lehrerfahrung

2009 Lipari RNA school

2005–2008 Mentor für Systembiologiedoktoranden.

2000–2002 Bioinformatikkurse für Doktoranden. Aufbau und Lehre von Kursen für Algorithmen und Methoden zum Proteinsequenz-Alignment, Mustersuche, 3D-Proteinstrukturbestimmung und Microarray Analyse.

1994–2000 Tutor für Grundstudium Mathematik, University of Manchester, UK.

Vortragsliste (nur 2009/2010)

2010 ISMB Konferenz, Highlights Track
2010 IBM, New York, USA. Invited seminar series
2010 Applied Pharmaceutical Technology Konferenz
2010 Humboldt Universität, Molecular Systems Biology Seminarreihe
2010 Les Houches, Frankreich. Title
2010 Broad Institute, MIT, Boston, USA Seminarreihe
2010 NERD club, Harvard University,
2010 Universität Turin, Italien. Zentrum für multidisziplinäre Forschung in Optimisierung und Interferenz
2009 Toronto University, Canada. Donnelly Center Seminarreihe.
2009 Recomb Regulatory Genomics
2009 Keystone 'Cancer and microRNAs'. Colorado, USA

2009 CNIO, Madrid, Spanien.
2009 Liparin School, Sizilien, Italien
2009 Weizmann Institute Seminarreihe. Invited speaker.
2009 Gotenberg University. Invited speaker, Colloquia.

BERLIN, DEN 10.05.2010

DEBORA MARKS

9   Publikationsliste

Debora S. Marks
Doktorarbeitstitel: Computational Biology of microRNAs and siRNAs

1. Larsson E, Sander C and **Marks DS**, *mRNA half-life limits microRNA and siRNA activity*. In review, May 2010, Science.

2. Jacobson A, **Marks DS*** Krogh A*. *Discovery of RNA binding protein signatures which antagonize and synergize with microRNA regulation.* Accepted May 2010. Genome Research * Joint

3. Arvey A, Larsson E, Sander C, Leslie CS, **Marks DS**. *Target mRNA abundance dilutes microRNA and siRNA activity.* Mol Syst Biol. 2010 Apr 20;6:363. PubMed PMID: 20404830.

4. Khan AA, Betel D, Miller ML, Sander C, Leslie CS, **Marks DS**. *Transfection of small RNAs globally perturbs gene regulation by endogenous microRNAs.* Nat Biotechnol. 2009 Jun;27(6):549-55. Erratum in: Nat Biotechnol. 2009 Jul;27(7):671. PubMed PMID: 19465925; PubMed Central PMCID: PMC2782465.

5. Betel D, Wilson M, Gabow A, **Marks DS**, Sander C. *The microRNA. org resource: targets and expression.* Nucleic Acids Res. 2008 Jan;36(Database issue):D149-53. Epub 2007 Dec 23. PubMed PMID: 18158296; PubMed Central PMCID: PMC2238905.

6. Betel D, Sheridan R, **Marks DS,** Sander C. *Computational analysis of mouse piRNA sequence and biogenesis.* PLoS Comput Biol. 2007 Nov;3(11):e222. Epub 2007 Sep 28. PubMed PMID: 17997596; PubMed Central PMCID: PMC2065894.

7. Chang J, Nicolas E, **Marks D**, Sander C, Lerro A, Buendia MA, Xu C, Mason WS, Moloshok T, Bort R, Zaret KS, Taylor JM. *miR-122, a mammalian liver-specific microRNA, is processed from hcr mRNA and may downregulate the high affinity cationic amino acid transporter CAT-1.* RNA Biol. 2004 Jul;1(2):106-13. Epub 2004 Jul 1. PubMed PMID: 17179747.

8. John B, Sander C, **Marks DS**. *Prediction of human microrna targets.* Methods Mol Biol. 2006;342:101-13. Review. PubMed PMID: 16957370.

9. Monticelli S, Ansel KM, Xiao C, Socci ND, Krichevsky AM, Thai TH, Rajewsky N, **Marks DS**, Sander C, Rajewsky K, Rao A, Kosik KS. *MicroRNA profiling of the murine hematopoietic system.* Genome Biol. 2005;6(8):R71. Epub 2005 Aug 1. PubMed PMID: 16086853; PubMed Central PMCID: PMC1273638.

10. Leaman D, Chen PY, Fak J, Yalcin A, Pearce M, Unnerstall U, **Marks DS**, Sander C, Tuschl T, Gaul U. *Antisense-mediated depletion reveals essential and specific functions of microRNAs in Drosophila development.* Cell. 2005 Jul 1;121(7):1097-108. PubMed PMID: 15989958.

11. Chen PY, Manninga H, Slanchev K, Chien M, Russo JJ, Ju J, Sheridan R, John B, **Marks DS**, Gaidatzis D, Sander C, Zavolan M, Tuschl T. *The developmental miRNA profiles of zebrafish as determined by small RNA cloning.* Genes Dev. 2005 Jun 1;19(11):1288-93. PubMed PMID: 15937218; PubMed Central PMCID: PMC1142552.

12. John B, Enright AJ, Aravin A, Tuschl T, Sander C, **Marks DS**. *Human MicroRNA targets.* PLoS Biol. 2004 Nov;2(11):e363. Epub 2004 Oct 5e264. PubMed PMID: 15502875; PubMed Central PMCID: PMC521178.

13. Pfeffer S, Zavolan M, Grässer FA, Chien M, Russo JJ, Ju J, John B, Enright AJ, **Marks D**, Sander C, Tuschl T. *Identification of virus-encoded microRNAs.* Science. 2004 Apr 30;304(5671):734-6. PubMed PMID: 15118162.

14. Enright AJ, John B, Gaul U, Tuschl T, Sander C, **Marks DS**. MicroRNA targets in *Drosophila.* Genome Biol. 2003;5(1):R1. Epub 2003 Dec 12. PubMed PMID: 14709173; PubMed Central PMCID: PMC395733. 13: Aravin AA, Lagos-Quintana M, Yalcin A, Zavolan M, Marks D, Snyder B, Gaasterland T, Meyer J, Tuschl T. *The small RNA profile during Drosophila melanogaster development.* Dev Cell. 2003 Aug;5(2):337-50. PubMed PMID: 12919683.

15. Cole C, **Marks DS**, Jaffar M, Stratford IJ, Douglas KT, Freeman S. *A similarity model for the human angiogenic factor, thymidine phosphorylase/platelet derived-endothelial cell growth factor.* Anticancer Drug Des. 1999 Oct;14(5):411-20. PubMed PMID: 10766296.

16. **Marks DS,** Gregory CA, Wallis GA, Brass A, Kadler KE, Boot-Handford RP. *Metaphyseal chondrodysplasia type Schmid mutations are predicted to occur in two distinct three-dimensional clusters within type X collagen NC1 domains that retain the ability to trimerize.* J Biol Chem. 1999 Feb 5;274(6):3632-41. PubMed PMID: 9920912.

17. Brockwell DJ, **Marks DS**, Barber J. *Structural investigations of kirromycin bound to bacterial elongation factor Tu by NMR and molecular dynamics.* Biochem Soc Trans. 1997 Nov;25(4):S612. PubMed PMID: 9450040.

18. Bichenkova EV*, **Marks DS***, Lokhov SG, Dobrikov MI, Vlassov VV, Douglas KT. *Structural studies by high-field NMR spectroscopy of a binary-addressed complementary oligonucleotide system juxtaposing pyrene and perfluoro-azide units.* J Biomol Struct Dyn. 1997 Oct;15(2):307-20. PubMed PMID: 9399157. *Joint

19. Bichenkova EV*, **Marks DS***, Lokhov SG, Dobrikov MI, Vlassov VV, Douglas KT. *Structural studies by high-field NMR spectroscopy of a binary-addressed complementary oligonucleotide system juxtaposing pyrene and perfluoro-azide units.* J Biomol Struct Dyn. 1997 Oct;15(2):307-20. PubMed PMID: 9399157 * Joint

Berlin, den 10.05.2010

Debora Marks